Managing Predators and Understanding Fear

Mark Coats

AF797193

Dedication

To any Rancher searching for solutions to predatory pressures.

Contents

Preface

Predators are a risk to any Ranching Operation. The risk that they present is to our stock. Such risks are not only depredations but the unsettled well-being of the stock. Predators affect performance as well as demeanor, breeding, and the inefficient use of the natural resources.

Predator Awareness was developed to mitigate such risks.

By addressing both sides of the predator vs. livestock encounter through instinctual responses, we interject our intentions without encountering the predators personally. Providing an effective Safe Zone for our stock

Chapter One

A Basic Understanding Brief

Predator Awareness is about *PRO-ACTIVE STEWARDSHIP* that insures *Natural Sustainability*. By addressing both sides of the *predator vs. livestock confrontations* we interject our intentions without our presence or engaging the predators *personally. It is a non-lethal set of solutions that can effectively deter confrontations.*

First is Scenting, placing a strange and unfamiliar scent within the natural environment which creates an uneasiness. Such a non-understandable scent draws forward an instinctual response, which detours presence. Communicating threw a predator's most powerful sensory perception, its o'factory system. The unfamiliar stimulus reaches *the instinct of self-preservation. Engaging a response of fear.*

In order for these *scents to be effective*, they must present an *unknown*. That *unknown factor* is the *scent deterrent's effectiveness*. Don't scent with a pine-scented manufactured scent if you're in the proximity of pines. Awareness of your surroundings is key in choosing an effective *scent deterrent*. It must be an unfamiliar stimulus; an example would be a citrus scent in a pine environment.

Many who place cameras will use an attracting scent; we are trying to deter, not attract. If scenting is done properly, cameras will not show any predator photos. You are Deterring presence.

Products that have *beef by-products incorporated into their mix should be avoided.* We humans can't distinguish those incorporated scents, but the sensitivity of predators may. Some examples are Downy fabric softener and many shampoos and conditioners. Sugars often have beef byproducts and are an attractant.

Reading labels is a challenge, they don't say *beef byproducts*. They list such things as *dimethyl ammonia chloride, Panthenol, Amino Acids or vitamin B*. The list of byproducts is large, if you have a question, I suggest googling it, before buying it and inadvertently placing an attractant.

Your grocery stores cleaning or detergent isle's are an easy location to find such manmade manufactured scents.

Placing scent requires thought. It's not just simply throwing something out there. It's about strategically placing an unknown and continuing to present an ever changing unknown.

If your smeller isn't sensitive then ask someone who's smeller is, to accompany you when purchasing scent for deterring. Effective scents in their concentrated form are so powerful that you can smell them through their packaging.

When you present these deterrents, we don't want any

understanding for the predators. *They cannot contact it. Contact provides understanding.* Elevate its position to avoid contact.

Contact of the scent deterrent as well as human scent only heightens the predator's elusiveness. Providing the predators with understanding. That understanding can not engage an instinctual response. Understanding *removes fear* and promotes comfort.

To our Stock

Second, by re-establishing a faded instinctual response within our stock, the Defensive Posture of the Herd Group, we reestablish a comfortable and effective response to a predator's presence.

It is a skill that Ranchers can practice that is a low cost and effective solution to any predators' pressures. The skill requires Stockmanship as well as an understanding of an animal's behavioral tendencies.

In my years of asking many *wildlife biologists*, I seem to repeatedly get the same response. My question is how many investigations have you found a wolf attacking the head of a cow? The common response is not many. I then ask, 5? They're response is curious, then I go down the list 3,2,1? "Well, I have never found one," is the common response. During an investigation the investigators will skin the hind quarters, or the forelegs arm pit, looking for signs of a wolf attack.

That overly abundant fact is not simply coincidence. It is a steadfast behavioral characteristic of the canine species. Canines require the chase sequence, putting them into a position of control. The *Standing Solution interrupts* the requirement of the *chase.*

By discouraging individualization and promoting *herd awareness* while posturing with the *Standing Solution*, we interrupt that predator prey relationship. By developing the cattle's *draw of the eye* we engage a herd posture that creates an uneasiness for predators.

By rekindling a faded instinctual response that's naturally embedded within all herd animals, we promote and reestablish the *Defensive posture of the herd group.*

Thirdly, it is about routinely monitoring and reading and understanding signs of a predator's presence as well as changing and presenting a new scent as to not allow the predators to become desensitized.

These methods have shown a significant reduction in predatory conflicts.

A Chance Occurrence

Many believe that a predator's presence is simply an occurrence of chance, much like the role of the dice, or the drop of the ball at the roulette table. Perhaps, but what if you filled specific holes on the roulette wheel or used a loaded pair of dice. The chance still exists, but for the most part it would diminish the likelihood of such an occurrence.

When we address animal interactions, it's much like tilting the odds in our favor. Although the chance still remains, the probability of a *chance occurrence* is greatly reduced.

Purpose and Design

Predators are a challenge to any Ranching operation. Many deterrents fit smaller ranges, but few are effective on large ranges such as Forestry Permits. *Predator Awareness* was created to fit that specific need.

With Fires and fuel loads being forefront on many minds and grazing being an integral part of that fuel load reduction effort. Grazing will continue to be part of an effective lower canopy fuel load management practice.

Predator Awareness is simply using an animal's own *behavioral characteristics* to your own personal benefit.

(Mismanagement and increasing fuel loads destroy wildlife and their habitats)

As urban areas continue to expand to create new and vast suburban communities, the agriculture production base continues to shrink.

As the worlds population continues to expand and the rising population requires more and more food. The problem is that the suburban expansion continues to cement over the most productive farm ground. These lands are the premium grounds used in the growing of row-crops, such as vegetables.

Grain and cereal crops can be grown throughout the Midwest, although much of those crop yields have been mandated for fuel. That mandate has taken those yields out of the food chain.

Row crops are only growable in a Mediterranean climate, along with an adequate water supply. Making those areas a finite and shrinking productive area. These areas are a very comfortable place to live. These desirable living conditions are doing away with premium vegetable producing lands.

The areas that will have to produce a higher qaulity protein source for the expanding population will be the rougher non-farmable lands that are the majority of the vast landscapes within North America.

These marginal lands are the hills, mountains and deserts. The product that they produce is Beef. The cow eats that sage brush or the grass or other palatable browse and returns a premium product. Nourishing the ever-increasing masses.

Although these increasing human populations want to eat, they also want their wild lands to enjoy, with all of the native wildlife as well as the many introduced non-native species.

Protections for many of these species are already in place. What that presents to producers is working within those laws. Making the predators presence a mandatory requirement.

One can either break the law or suffer the economic losses that the predators bring with them. The only other alternative is to work with the *behavioral characteristics* of the animals.

A Beneficial response

By providing an alternative response to *fear* for our stock, rather than flight, what we accomplish, is first an answer to the economic challenges that the predators bring with them.

Secondly, *we address our consumers' concerns about our Stewardship* and whether the Livestock Industry acts responsibly in the care of the land and the environment.

Thirdly, we begin a conversation that substantiates the need for a successful Beef Industry. By producing a sustainable and desirable product as well as the benefits to the environment itself, such as fuel load reduction, as well as habitat improvements. Such as the added drinking water that benefits the wildlife as well as the stock.

The stocks grazing as well as browsing, creates hoof prints that incorporate seeds into the soil, as well as creating a soil texture that holds moisture. That track and its agitation helps avoid erosion as well as creating a small reservoir to help the incorporated seed in its germination. The track breaks the soil, preventing it from becoming crusted, developing a healthier and higher qaulity sub soil, as well as sustaining a higher soil moisture profile, benefiting the entire ecosystem. Increasing plant growth, improving wildlife habitat, sequestering carbon, increasing oxygen production, mitigating fire risk and stopping the carbon eruptions that those fires generate.

(Prescribed burn on the Tule Lake wildlife refuge)

Grazing livestock would turn that unused plant material into a usable food supply. While promoting carbon sequestration rather than promoting carbon eruptions. Addressing and providing food to fight world hunger.

By producing a premium and desirable food product responsibly, we meet an ever-increasing demand for a high quality and affordable food supply for the ever-increasing world populations. Mitigating the steady loss of productive farm ground and its production.

Thought Confrontation

When we reflect on predators, we as producers automatically have thoughts of our stock's confrontations. Our thoughts go to our concerns first and foremost.

Whereas environmentalist's thoughts go in the direction of the admiration of the species they most admire.

Humans then confront each other because of their opinions based on their own opinion of what they believe the other side thinks.

This *thought confrontation* between humans has no meaning or any concern to the animals. Whether it's the predators or the prey, whatever side humans put their selves on is of no concern or meaning to anything other than the human opposition that you're opposing.

This opposition if one would like to refer to it as such, finds its way into society routinely. You can call it Politics or Bureaucracy or the Media, any definition would be correct, because they're all ways of addressing the other side rather than the actual problem.

Man is innovative, whereas societies are subjective.

If one thinks of laws or regulations, those laws and regulations actually aren't about animals. They're about people. I realize that many are stewing right now and want to yell out, it's about protecting the animals!

I understand your point, but it's about stopping a human action that is a perceived threat to the animals in some way. The perception is that it jeopardizes some facet of the animal's well-being, and that facet is always a human interaction of some sort. Often a perception may not be fact, it can simply be one's misguided interpretation of a situation.

An animal has no realization of anything man deems lawful or unlawful, they only have their behavioral characteristics.

In actuality, the animal's adaptability is why the existing species are still here. I realize that mankind has intentionally affected specific species such as the buffalo, but man's effort was a targeted assault on them. Most of the animals lost, have vague analysis to the reasoning behind the loss.

Realizing other reasons for such losses, such as the Avian Flu or other bacterial or viral infections are certainly a possibility. The technology of the times just wasn't there, to have any definitive answer.

I realize many reading this will think about the dinosaurs. The dinosaurs were wiped out because of the Ice Age. There is a lot of geological evidence that proves the Ice Age was fact. But when you think about most of the dinosaurs being reptilian, cold blooded. The theory of the Ice Age being a rapid event, because of some outside force such as a meteor strike, gives an understandable reasoning to the dinosaur's extinction.

But, if you can agree that the Ice Age happened and that rapid cooling was the demise of many cold-blooded species, that poses a question, doesn't it?

That question is then, if the Arctic was grazing lands and the Sacramento and San Joaquin valleys were bays and cold-blooded species ruled the earth what was earth's natural temperature before the meteor struck it?

Global Warming is a social anxiety at the moment. But, if one thinks about the natural world being a much warmer place and warm-blooded animals fading away and new reptilian species emerging, it makes one think of a much larger cycle, doesn't it?

Just to give some perspective to the anxiety global warming conjures up, in 1901 the thought was mankind would be buried by horse manure, motivating and guiding the industrial revolution.

If warming is a pre-destined conclusion and possibly the Ice and Cooler temperatures will only be a small part of the futures history. Then anything man does is trivial and pointless to say the least.

Possibly the cycle we are in, is still coming out of that Ice Age and returning to those elevated temperatures that existed before that meteorite changed the evolution of this planet.

Possibly, global warming is about power and financial gains.

Agendas

The thought of writing a law that is intended towards an animal is silly, the laws are about people. The facts that are presented in creating these laws and regulations are about people, therefore they're filled with agendas as well as *profiteering*.

We don't correspond verbally or with the written word with animals. When we correspond with them it's through interactions and *behavioral characteristics*.

If the concern is politics or regulation, that's OK, one just has to realize that those efforts are about people. Those efforts are intertwined with all of the pitfalls of mankind and his quests. The human factor in the equation is actually the confrontation.

The animals only have their *behavioral characteristics*.

Predator Awareness is about stepping away from that human entanglement and addressing a species *behavioral tendencies.*

Economic effects

There are many different deterrents when dealing with predators. An equal understanding across the board is that predators affect the bottom line. When I say the bottom line, I'm not just referring to Ranchers but to a County, State, or a federal budget. Predators cost more, whether it's deterring, management, restitution, monitoring, or even studies.

The economical impact reaches everyone. When the Politicians are working on budgets, they have constituents that must be answered too. Unfortunately, the masses don't consider those expenses that important. Leaving resource shortages for predators just about everywhere you look.

Complacency

One of the situations that has surprised me, as I fell into this category myself. I had become *complacent* with my predatory losses. Until the famous wolf OR7, who's journey to California was followed by so many, passed through my ranch.

I'm not sure if complacency is the correct word a better maybe distracted. Challenges presented in the daily routine of production agriculture just occupy much of one's time.

My beliefs had fallen into the mindset that predatory losses were just an *expense of doing business*. But when the Wolf entered into the picture my concerns became elevated. The public turmoil that the Wolf brings with it, elevated my anxiety.

Because of the increased drama that the wolf brings, it perpetuated my efforts to resolve my losses and realize that any loss to any species is unacceptable.

Responsibility

When were talking about economics, it shouldn't matter what species created the loss. The loss isn't any different because of what species was responsible, the loss is the same wether it's a coyote, lion, bear or a wolf.

Often, those who are developing any plan for restitution focus on a specific species that was to blame. In actuality, the *loss* is the reason you are seeking restitution.

The question then becomes, what's the reason for restitution? Is it the financial *loss,* or the importance of the knowledge of what species was responsible?

If the loss was because of a coyote, or bear, or any other reason than the wolf, such efforts as seeking restitution are pointless.

Then, the effort of *deterring presence and deterring conflict* maybe a better avenue than waiting for the possibility of the chance of a restitution settlement.

Undervalued Restitutions

Often, when a restitution is rewarded, it's at the current market value. As a cow-calf producer, I would like to point out that the calf is how we pay our yearly expenses. The budget is expecting that calf's income at that estimated market value.

The expenses are paid first, and we then receive the market value when we sell the calf at the end of the production year. The estimated value of the calf is what the annual budget is developed from.

The loss of the pre-mature market value is a shortfall, that must be made up from the other marketable calf's average. Added to that equation are the costs that are associated with the loss. That one calf was intended to pay its portion of bull investment, pharmaceuticals, trucking, mortgage and rents, winter feed costs, labor, insurance, cost of living, and reinvestment. Devaluing the entire calf crop. This picture shows a calf loss of a market value estimate of around $400. The investment to date would be closer to $900.

With the depredation, there are fewer calves to cover the expenses and costs. The estimated income is then reduced, and costs that have been paid are not adjustable. The remaining calves must now carry the added financial burden.

When a live calf sale happens, a hundred-dollar return would be considered acceptable.

That one depredation then took five additional calves returns with its loss. Then, by removing those calves' contributions towards the expenses, you further compound such a loss.

A restitution should be based on a three-year average of that ranches returns at the intended marketing dates.

If a breeding female is measured into the loss, such losses are then only compounded and expanded into three extended years, until that productive mother is replaced. Incurring the calf-hood retention, which was intended income. Also, the devalued first calf's return, as well as the cost of caring for those three years, as well as the breeding costs.

The actual lost mother cow value is variable and depends on her age and how many calves were left in her productive years, as well as her salvage value. Most Ranchers realize 10+ calves for a cow's lifetime. Requiring the replacement cow's entire life's production to make up for the lost mother's accrued losses as well as the replacement costs to get back to your financial position prior to the depredation.

When we're talking about breeding stock, we also must consider ancillary costs, such as open cows (cows who didn't breed back). Predators do affect gains as well as conception rates.

If the *predator's presence* rather than *a depredation* was to blame for financial loss, such as an open cow. Then there is no avenue for restitution. But that open cow's expenses and loss are just as substantial.

Costs versus Cost

One of the comments I hear is that the time in deterring predators is expensive and requires too much effort. Losses are expensive. But predators create expenses wether they incur a loss or not, whether it's the deterring effort or the many other ancillary costs presented by the predators.

As to my case, I believe that the added calves, 63 to date, over 13 years, help to mitigate those costs. I realize that 4.84 calves per year in a business that works on such tight margins, as well as many other contributing factors, such as weather and market volatility, aren't a large gain, but it's still a gain, rather than a loss.

If one's concerns is the expense of deterring confrontations, one might weigh the saved calves into the equation as well as the benefits of good gains and breeding conceptions.

I realize that on paper, it's a challenge to count a nonoccurrence to the benefit of the costs of deterring. If you're weighing out the financial expense, you then have to look at, do the possible benefits outweigh the risks?

The challenge then becomes, after a successful deterring effort, the thought is, we *just don't have any predatory problems*, the budget is tight, *let's make some cuts. The successfulness is the effort's weakness when it comes to budgetary cuts.*

Self-preservation

The instinct of *Self Preservation* is within all living creatures. The response that the instinct engages is *fear. Fears* responses are *fight or flight.*

The response of *fear* is to a perceived threat. The responses of *fear* are intended to elude or confront that threat.

All animals have instincts that govern *fear*. Predators, as well as our stock.

Predators have developed skills that exploit *fear's* responses. In order to effectively manage a Solution, we must understand that a predator requires an individual.

That individualized position is the *Chase Sequence*, putting the predator into a non-confrontational position of control while escalating *fear's* response within the individualized victim.

Herd animals are *herd animals* because of predators. The

herd itself is yet another response to *fear*. The *herd* offers a solution to the alternative of individualization.

Although our domestic stock are herd animals, their *herd* awareness has been diminished from the lack of apex predators along with our management interactions.

The current social interests of promoting the populations of all of the predatory species pose concerns for Ranchers.

These increased predatory populations will have a regulatory action that will be directed towards Ranchers.

A Rancher's responses and actions will determine how severe those regulatory responses will be.

Instincts

Instincts are a part of the whole package; they're not removable or modifiable, but they can fade with nonuse.

Our Stockmanship skills and applied pressures can reinstall this faded instinctual response of the *Defensive Posture of the Herd Group.*

By simulating a predator's pressures with the aid of *a pseudo-predator*, we can re-install such a response, which then becomes the solution to predatory pressures.

The *pseudo-predators* don't *physically assault* the stock, but rather use their speed and elusiveness to confuse the herd group. In the first few sessions, the stock are *fearful*, and the speed and circling of the group by the *pseudo predators* is what creates their *fear*.

The stock soon engage the Standing Solution, which is the response we are seeking.

When we practice this response, we use a basic training skill of desensitization. Desensitization removes movement; with each session, the stock become more responsive and develop a *calm resolve* when they are pressured.

Preparing for the Response of Fear

(Training)

In the attempt to explain and understand *fear*, I must relate back to humans to illustrate my point. *Fear* is a response; think of some occupation that encounters a *fearful situation* routinely. I am going to point out for this analogy, Firemen.

Firemen simply don't show up to a burning building unprepared, wondering and discussing, what do I do? They have spent a lot of time preparing for this serious situation that they are now presented with.

Their *training* has developed their required responses. They have practiced these dangerous situations repeatedly; those practice sessions have instilled a successful performance to this task at hand.

Their training has also given them confidence in their ability, and that confidence promotes a *calm resolve* to step up and do their part to protect not only property but also lives.

That performance isn't an occurrence of chance, but a *trained and coordinated* response. The effort of *Predator Awareness* is very similar to the pre-conditioning of first responders. Instilling a *solution* or *response* to a situation of 'just in case.'

Taking that *trained* performance of the Firemen as an example of our stock effectively managing a predator's pressure, may be a reach for some. But the *training* removes *fear* and instills a response, which is my point.

Training a response?

I have often heard from many, "You can't train cattle." Actually, history shows us that's not true. Oxen have been used in logging and farming as well as transportation. Only the Industrial Revolution ended their role as beasts of burden. This brief acknowledgment tells us that an effort will be successful in its outcome if done properly.

As to that same thought of training, we are not creating a new trainable task; we are simply rekindling a *faded instinctual response* that is naturally embedded within our stock.

If one were to attempt the effort haphazardly, it would simply play into the predator's playbook of ineffective training and promote *fear*, which a predator will then exploit.

The *Defensive Posture of the Herd Group* and the *Standing Solution* are more than a group of cattle. It's a practiced response to a predator's pressure. It develops an awareness to not become individualized.

In comparison, the *Firemen* are a prepared group, whereas a group of random individuals are not.

Denying Opportunity

The Solution gives our stock a safe zone, so to speak. But it also challenges the predator's *self-preservation instinct*.

The predators that require an individual are *denied* that *opportunity*. Challenging the predator to confront the *defensive posture* of the herd group.

That challenge, which is the *Standing Solution*, generates doubt; doubt is the first step in the engagement of *fear*. The doubt is a concern of the engagement, only leaving flight as a response. Simply asking the predators to move on.

Re-directing *fear* from our stock and presenting it back to the predators.

When we *rekindle* this response, we not only present a calm resolve for our stock by replacing *fear* with the *Standing Solution*, but we create an instinctual response within the predator as well.

Drawing on instinctual responses from both sides of the *predator vs. livestock* encounter.

All the *Solutions* of *Predator Awareness*, whether it's *Scenting* or the *Standing Solution*, both are about effectively managing the *instinct of self-preservation* and its response to *fear*.

These methods are effective in preventing losses from predators. They address the livestock as well as a specific range. They are most effective when there is no human contact with the predators. These are Pro-Active Stewardship techniques that effectively mitigate predatory risks.

Chapter Two

Selecting a Deterrent

Often, when we make a selection, it's not based on an analytical conclusion but rather a preference for what we like or dislike. These are everyday choices that shape our selection methods in a very broad way.

Whether choosing an itinerary, a dessert, or even a predatory deterrent, it often comes down to a choice of what one likes the best or dislikes the most.

I remember stopping at a BLM feedlot where the *Adopt a Mustang* program was happening. My wife and I, as well as others, were viewing the large selection of adoptable candidates.

It was not uncommon to hear comments about "Oh, I like that one," which would be followed by its color or looks, such as its beautiful mane…or I like…its eyes".

The liking or disliking of something is often an emotional thought rather than a reasoned conclusion.

I realize that everyone has their preferences, as do I, and I have to say I don't like white feet on my horses.

In choosing a horse, for me, it's not about looks, but rather, what individual will spend its lifetime in a good working partnership with me.

When I make a decision on an animal related subject, I always begin with the long-term objective. I prioritize what the desired outcome is. I realize that the effort is a long-term analysis for such a quick response. But, through the years and plenty of misdirected choices, I pretty much realize what my needs are.

My long-term objective is remaining in business, adapting to each new challenge, including predators and the challenges they present.

I realize from being in the Ranching business for my lifetime, there are things that I would like to see or accomplish. But the Ranching businesses streets are not paved with gold, although many might think they are.

Budgetary issues are first and foremost in all decisions. What one wants is often what one can't afford, or simply can't do anything about, such as more rain or less of it. Wishing or hoping or even praying doesn't often change the situation. Whatever the problem is, the long-term strategy should be one that engages long-term solutions.

Worrying about predators won't change the fact that strategies have been put in place by society, that are increasing the populations of all of the predatory species.

A reactionary response would be to wish them away or just eliminate the problem immediately, such as just "Shoot them." Well, if you look at historical data, that method of control and management hasn't been too successful to date. You can look to Coyotes; we have had that "just shoot them" management practice for a long time and yet the Coyote is still a management concern for a lot of Ranches.

The Wolves' territory continues to expand, and I know that the idea of "Shoot Shovel and Shut Up" is not a new concept.

Even where States have legalized hunting of the Wolf, Game Departments have realized lower-than-expected takes. Distinctively discounting hunting as a 100% effective population control and management solution.

While many seek solutions through legislation or eradication, my beliefs are that those reactionary responses have been shown to be ineffective in mitigating the risk to our stock.

Each of the predatory species' populations has continued to expand and spread over the landscape. The protections that have been established and the elimination of effective controls have been very successful.

For the activists, it should be a celebration of success, but the response is a continued effort to remove the Ranching Operations from the private and public landscape.

From previous actions, the activist's record shows little concern or benefit to whatever they use as a prop for their real agenda of reshaping society and gaining control over it. The highlighted animal is just a disguise to instate their political agenda.

Realizing your needs rather than your wants and making a commitment to the long-term objective may include changes in historic practices. Just as changing to market preferences, choosing solutions that fit and promote a sustainable conclusion, requires serious consideration if remaining in production is the goal.

Just as when droughts occur, they challenge what we do. They present problems that aren't within our normal operating parameters. The stock, as well as our ranges, are stressed, and our normal carrying capacities have to be reduced or spread out further. The long-term goal must be remaining in business rather than trying to perform normally, no matter what the cost.

That simple reality of immediate management decisions over a wait-and-see approach, will be a continuing story of success over failure each and every time.

Predators are much like the drought; they present unique challenges to our management practices. We can focus on the predators, spending a lot of money and effort in that direction, or we could focus on our stock and deterring presence affordably.

Although the word *prejudice* relates to human interaction, such as religion, race, lifestyle, or ethnicity| it can also be a tool in passing up a well-thought-out position on an animal's behavioral tendencies and a long-term sustainable solution. The prejudice would be to the change of traditional practices.

"We've always done it this way" ....(fill in the blanks).

To that point, just as society is changing and their political positions to *the old ways* are also imposing more restrictions on production. Adapting to evolving pressures, whether political, regulatory, or predatory, all require attention rather than a wait-and-see approach.

A behavioral characteristic of humanities progress is simply to leave the old ways behind.

In my opinion, by leading and showcasing our own management practices, we address and resolve many of society's perceived problems with our production methods. By adapting and adjusting the old ways and presenting to our consumers a new, more productive, and beneficial role in proactive Stewardship, we present an affirmative, proactive response.

Leading and developing these cutting-edge solutions while promoting and creating a very marketable position for sustainable agriculture.

It not only gives us an opportunity in developing new support and gaining a larger portion of the marketplace. Satisfying the general public's concerns about working with the land and its wildlife while producing a proactive and sustainable product.

Public interactions

Society continues to change, and its connection to food production and the understanding of the use of our natural resources appears to have been lost to the general public. That may be from the steady assault, not against agriculture, but rather to reality. Many people believe that nature is just one big happy lovefest and our products just self-generate themselves within the store; in reality, nothing could be further from the truth.

By engaging the public and expounding our proactive efforts, we change the argument to a conversation.

In the natural world, nature creates harmony *by death.* A plant dies, and it is then incorporated into the soil, enhancing the soil, which then feeds other plants. An animal dies or is killed, and from its death, it feeds others. One dies and benefits another's life; that is, *"The cycle of life."* It's reality.

Through the years of "Can't we all just get along mentality," we have come to this effort of making all of God's creatures happy. Changing the natural order by passing laws and protecting a preferred species over another for its political correctness of the moment. Only promoting *financial gains and a feel-good moment* for human's and accomplishing nothing for the natural world.

These efforts to reestablish or Co-exist, or whatever happens to be the *flavor of the day,* only point out the futility of modern society's efforts to re-construct the natural world.

The effort is a guise for profit and to control another portion of society's behaviors or gain an assemblage of power over another's lifestyle.

Unfortunately, we who are involved in the production of food that feeds these masses have become the objective of their futility. But, in reality, the effort maybe a much larger quest for power, such as Communism and *Karl Marx's* position that Agriculture is the key to Capitalism.

Our roots run deep.

In my grandfather's time, most Americans were directly involved with Agriculture. Since that time, for those who are in production Agriculture, their numbers have continually declined. Yet the worlds population has continued to expand and hunger across the spectrum has continued to diminish.

The fact is whatever the challenge, the Agricultural community has continued to meet society's needs. The Agricultural community, although shrinking, still feed the masses and those masses still conspire to collapse those agricultural producers' production.

The effort then becomes one of surviving these challenges as well as the ever-increasing predator populations. By understanding animal behavioral characteristics, rather than challenging them or engaging their human proponents in nonsensical rhetoric, we avoid conflict for our stock as well as ourselves.

So, the decisions we can control may not be our first choice, but rather the best scenario for the moment and the budget. Allowing us to proceed on our journey to that long-term goal.

"Life, liberty and the Pursuit of Happiness" **Thomas Jefferson.**

Chapter Three

The Natural world

The *natural world* is always at work, seeking its equilibrium. It's much like a set of glasses linked together at the bottom of each glass with a tube. Pouring water into one will allow the water to eventually be at an even level in each glass.

Pouring the water too fast into an individual glass could be thought of as something like a flash flood, exceeding the glass's boundaries. Much like a river running past its banks.

In nature, extremes do happen, but the average is the balance over a measured time or distance.

If one thinks about one glass filling first and then as time goes on, the water level becomes even within each glass. If one were to measure the brief high water level against the other's empty status, one would see an extreme ratio for that moment.

A response to the extreme ratio, that wasn't an actual problem, but was just an anomaly. One could present any solution or agenda, and the results would be the same. What's required is just time to allow the anomaly to balance, much like the water in the glasses seeking their equal level.

That brief summation is a reactionary response to an anomaly that was by all means unavoidable. It was simply a matter of chance that you poured the water too fast, or the patterns aligned by coincidence to create a severe weather pattern.

But, if one looks over a larger timeline, one will see a more accurate measurement.

When we talk about deterring predators, we must understand that the only effort that will be 100% effective will be to not present the opportunity at all. The Solution then, which can be followed throughout history, is to quit production and do something else.

But, if one realizes that anomalies do happen, such a deterrent is effective when measured over the course of time. When such an anomaly occurs, one should have solace in the effectiveness and weigh the gains versus losses to one's bottom line.

The water in the glass is an example and a thought of a way to lay out the example of balance. I realize that there are many whose occupation is water and would have a lot more to add.

The natural world is like the water seeking its level. If one thinks of another example, one could think of a simple scale balancing.

When it comes to animals within *the natural world*, we have to think about their density as well as their food and water supply. Many of Mankind's improvements within nature have greatly improved the quality and abundance of nature's resources.

Looking at one brief anomaly in an instant of time may serve a purpose, but the purpose may have alternative objectives.

California's perspectives

Many rank the bovine as a scourge or a stain on the West Coast, their historic numbers would have been at their maximum through the mid-1500's through the 1700's. At that time, the California Dollar was a steer hide, and the product was the hide and tallow trade. During its peak cattle inventory, California was a part of Spain.

The entire world sought and demanded these California products. The meat was simply a byproduct. Early on in California's history, it was an active agricultural supplier in the world economy.

I often hear comments about it being a world market and that we shouldn't exclude ourselves from that market. I understand, but if you look at history, we have been a world market since very early on.

The California 'hide and tallow trade' attests to that. Names like Marco Polo, whose ventures broadened the world's landscape, or Christopher Columbus, who was seeking a trade route to India, or empires such as the Roman, Greeks, and even the demands from the Orient and its silk trade. They all addressed the world market.

Long before California was a State within the United States of America, the Industry's numbers far exceeded anything that the

State of California would carry.

The key factor in the dismantling of the Rancho's was the discovery of gold and the Gold Rush of 1849. Those miners' masses had to eat. Through time and the urbanization of the West, the Cattle numbers have been in a steady decline.

The Cattle that so many attest to as detrimental, have been a functioning piece of the West Coast landscape for over 400+ years. Their density was the reason for the location of the Missions. Besides the Catholic Church, the Missions housed Spanish garrisons that collected taxes and protected the stock.

The lands that so many wish to protect, were actually a culmination of the Ranching Industry. The wildlife that thrived, only dwindled without such support from the Rancho's. The continued expansion and urbanization of California and the urban sprawl only degraded the wildlife and their habitats.

One doesn't have to use much imagination to think of what was better for the natural world, cattle and horses, or the current urbanization of so much of California. Such as the LA Basin and the population from San Diego to Santa Barbara.

Many believe that man's incursion has damaged nature's capacity. When you look at modern urban areas, it is certainly a challenge to find much left of the natural landscape. Yet, urban populations point to agriculture as well as the cows as the destruction of the Natural World.

Being involved in California agriculture, I have seen the many benefits to the wild that man has developed, intentionally and accidentally.

The efforts to manage through regulation and bureaucracy, have only degraded the natural resources. The new and changing dynamics that are damaging the environment are man's efforts to save it.

A Social structure

I have been in the Sierra Nevada and Cascade ranges for most of my life and remember the political interest that began as conservationists, naturalists, and then ecology, and then it became environmentalism.

After the youth of the sixties began seeking their positions in life, society realized their unpreparedness and lack of productive skills. So, ecological issues were created to elevate their social standing as well as employment.

These environmental issues shaped political and social dynamics and have morphed into a line of endless regulations and stacks of required forms and yet more paperwork to support the former form.

These positions fit the apparatus of government structure very well as the position engaged endless studies of anomalies and solutions that were pointless, but filled the needed agenda, which was driven by social profiteering and the development of yet another desperately needed agency.

When one thinks about governmental policy, one doesn't have to look too far, to see inefficiencies and the never-ending required forms, simply to feed a never-ending flow of pointlessness.

These pointless endeavors are the need to manage the overabundance of non-productive workers. Providing efforts and funding for the problems that don't actually exist.

The point is that they use fear as their tool for promoting the need to address these anomalies.

Unfortunately, these policies have shaped into agenda's that focus on people rather than the actual environment. Actually, becoming ways to do away with, or promote a lifestyle or occupation of their choice, rather than any effort of restoration or benefit to *the Natural World.*

The desire of society leaving the old ways behind, put's the progressive agenda in a lead position. The actual performance is irrelevant as long as the old ways are left behind.

These environmental positions have become one of the most valuable and profitable positions within society today. The money that is generated in this new green direction is one of the largest segments in the world economy. Wishing it away, or endeavors to change it, just as with the water anomaly, is pointless and will just take time to balance.

Endeavors maybe better focused on surviving rather than rectifying or changing the existing problems.

The challenge becomes survival and adaptability and is an integral part of enduring and remaining in business.

If one looks back to the environmental attack on the logging industries, one might ask, why did some survive, and others didn't? I believe the answer was adaptability. Those who changed and adapted endured, those who stood defiantly and would not change, faded into the memories of times past.

Such are the times we are in today. Adjusting to the season at hand, rather than wishing for snow in July. The snow will come, but it will be here on its own schedule rather than ours.

The point is the same when we address predators and solutions, we can have that thought of wishing for something different, such as changing the laws or public opinion. But I believe that's the same as wishing for that snowstorm in July.

Innovation

One could possibly assume that my position would be to remain in the past and not to move on to new horizons. I myself don't believe that to be true. I have a great respect for new ideas and am constantly seeking better ways and opportunities to improve my practices.

Innovation is stifled by bureaucracy and burdensome regulations.

Innovation is the response to need. Whereas regulation is about filing a new form and social conformity.

Ignorance can be disguised as progress, if the intention is actually leaving the old ways behind for political gains or profiteering.

New designs

Electric Vehicles are a hot topic of the day. So, let's chat about that idea. Too often, one's opinion of anything new, draws on their political position or *Selective Judgement. Selective Judgement* is shaped by political or social bias.

Many times, people judge an issue to its conclusion before ever engaging in the thought of its value or practicality.

As to the Electric Vehicle, one of the most common oppositions is that are Electric Grid simply can't handle any more Electric Charging stations. Or that our current grid isn't meeting the challenges of the current demands. As well as removing, clean and efficient hydro-electric facilities.

Not many who understand electricity, devalue the power of electric motors versus the carbon combustion engine. Electric motors may even supply a greater horsepower in a quicker response time. The issue is where do we get the energy to charge the required fuel of electricity that would replace that fossil fuel requirement?

Plugging it into a generation plant that is generating electricity from the consumption of fossil fuel seems to be an inefficient process if the goal is a reduction of fossil fuel consumption.

The storage of fuel then becomes an issue. Can we store an adequate supply of fuel, which is electricity? Just as we do for that Carbon Combustion requirement.

We have all seen the huge tanks at the refineries or the huge water tanks for a town's water supply, and we certainly have all changed batteries in our flashlights. Storage capacity of electricity is certainly an area that would greatly increase electricity's usefulness.

Many say that new battery designs are gaining in capacity and efficiency while reducing the enormous weight burdens that are currently required. While also enhancing the speed and volume of flow during recharging. Making gains in effectively storing electricity.

A current downfall for the storage of electricity is the storage units seem to have a large expense and short effective use timeline, making them cost prohibitive.

When one thinks about the benefits of storage within our Electric Grid, that fact in itself would greatly reduce costs and promote sustainability. Promoting a constant supply and elevating man's qaulity of life.

I realize that many would say that we're certainly not there yet. I understand, but as a boy watching a black-and-white TV. I witnessed Neil Armstrong Coming down the ladder of Columbia and planting an American flag on the moon. I also remember the comments in the media of hoaxes, or those who said it didn't happen. My Grandfather told me it reminded him of the responses that the Wright brothers received after Kitty Hawk.

My grandfather lived in an amazing time when innovation must have seemed to be a daily occurrence. He witnessed the beginning of the supply grid of electricity, the change of the Industrial Revolution, the Phone, flight and so much more.

So, I realize that in today's modern technological society, innovation seems so far away, but if you really think about what you have witnessed, the innovation of a more proficient electrical storage system seems an obtainable goal, doesn't it?

So, if we get to an efficient battery and we can adequately store electricity, think of the possibilities. Let's go back to that

Electric Vehicle. A generator, is a producer of electric current.

An Automobile moves itself down the road on four tires. What if the four wheels were generators? Producing and generating electricity while we were doing our daily travels and then downloading that generated electricity into a storage containment system at home, which our home then functioned and operated on.

Making the vehicle more than a useful transportation device, but itself becoming the key to advancement to our quality of life and completely eliminating your need for fossil fuels and their expense as well as meeting that green agenda.

I have been told by many that I am too idealistic, my thoughts aren't about ideals but rather man's ability to resolve perceived problems. Mankind's history certainly has some bleak moments, but Man's innovations certainly have been amazing.

Man's ability to step up to the task at hand is certainly one of man's finest qualities. Whereas Man's hunger for social dominance maybe his worst attribute. That "do as I say" mentality is often applied, but seldomly rewarded.

Food Production

Our charge, or our goal as Agricultural producers, is the production of food and sustenance for society. This simple, but challenging task has many legs that support this endeavor.

First is the economic sustainability of the operation. Whether we grow a vegetable or produce a steak, that operation ceases to exist if economic sustainability isn't achieved. If the market is adequate, then we must take care of our natural resources in order for our operation to continue.

There have been bad players in the game of Agriculture, but they're few and far between. If you think about Agriculture, you realize that farmers and ranchers are a diminishing portion of society. Yet, the worlds population has continued to grow.

With the expansion of cities, and suburban areas, the removal of the use of much of our public lands, and the acquisition of new parks and reserves, the production base of Agriculture has, just as with its operators, diminished in total acres. Yet, until political agendas challenged the supply chain, the store shelves have been continuously full. Providing food and sustenance for the ever-increasing population. Attesting to the care and sustainability of the lands that these Agriculturalists manage.

The word conservation has been used by many in relaying beneficial qualities to the environment. The word conservation simply means non-use.

If you're looking towards politics, conservation is paying someone not to use their resource and allow that non-productive resource to be administrated by a bureaucrat.

If an expenditure is created with no beneficial results or returns, one could say that the example is a perfect definition of government.

I live in an area that is surrounded by refuges and preserves. Their original intention was for wildlife habitats. What they have become is an overgrown habitat of weeds and rodents as well as a tool in degrading water rights for farmers.

Now, after many years of mismanagement, they have realized it's the farm ground that the wildlife is drawn to. So, in their wisdom, they're farming these conservation reserves. So, the word conservation has taken on a new twist, rather than just non-use, to a definition of use for only a select few. I believe that meaning then translates to tyranny.

One might think of my position as opposing government, but It's not. I oppose that high-in-the-sky attitude of knowing better than "We the People".

An example would be a health issue, let's say appendicitis, a severe case. You come down with that infliction, do you rush to the hospital or seek a homeless person on the street?

I know your answer but let me just remind you that a homeless person is going through tough times and could sure use some money. Your appendicitis could surely help them out.

The example, although silly, is a comparison to the bureaucracy that spews out solutions for the benefit of the environment while risking nothing, no consequences, only that "Do as I Say" mentality.

A Solution, overcoming inefficiency.

I previously mentioned my admiration of Innovation rather than pointless solutions. I believe that within private industry, a player, whether an employee, supervisor or management, should have consequences for inadequate performance or non-beneficial outcomes. That action exists, it's called capitalism.

The need is a review of government agencies; they should be required to account for their actions or inefficiencies.

On the political side, that's already in place, it's called elections. But bureaucracy is a position till retirement. Nowhere else is there that kind of job security.

The problem is these agencies have no accountability. At present, their only accountability is the judicial branch. A suit must be brought and then a court decides if their regulations are legal.

A review board should be set up, not to review individuals, but rather the agencies policies and whether they are effective in their efforts. Then, if it is found that they're not effective, then further investigation moves forward to understand why and make the needed corrections.

Establishing an oversight and installing accountability.

If it's because of the lack of funding, then more money would be required. If that funding couldn't be secured, then modifications would be installed. But I would suspect that a governmental agency's performance would be linked to other reasons for inadequate actions rather than the lack of funding.

The accountability and performance reform act would be an effort to succeed in intention and modify and adapt an agency's performance. Providing a successful outcome to its intent.

This review board would be an appointment much like court justices. The reason for the appointment would be the removal of political influences and social pressures. Each member would be answerable to the committee as a whole. Their actions or inactions would be a testimony to their performance rather than their connections.

By positioning such a review of agencies, those agencies would endeavor to manage projects and streamline themselves. In effect, making them reach for and achieve their stated purpose. Rather than simply heaping yet more pointless piles of paperwork upon one another.

My thoughts on bureaucracy may seem to run off like a herd of wild mustangs running off into the sunset. But it has been my experience that the predators and the challenge's they present are nothing in comparison to what man's efforts to manage the occasional anomalies are.

My point is managing *fear* is not a new concept. The fact is, that many already use this solution for managing *fear*. Manipulating *fear* and using it for their benefit.

By creating and supporting *fear* because of an anomaly, the risk is low and the funding extreme, but in actuality, what was needed was just time to allow the anomaly to balance back out.

Predator Awareness is about managing *fear* responses and applying it to animals.

Instilling a response within our stock and promoting a calm and safe solution to a predator's pressure.

Chapter Four

Our Actions affect Their Responses

Our efforts, as well as our time are valuable. We don't want to put in such time and then turn around and disrupt such effort by engaging in practices that counteract the stock's responses as well as benefiting the predators.

In order to evaluate our actions, we have to examine our Stockmanship skills. That's about examining our interactions with animals and how the stock respond to us.

Movement is intentional, when you approach the stock, do the stock move as a response to your presence. Or did they respond to your *pressure* and move as a relief to that pressure, or did they simply become aware of your presence?

The difference is about discovering your stock's reactions to your actions. The livestock's responses are about the stock's previous experiences and their perception.

The difference between those interactions is whether your interactions are based on a *work-with* approach or a *do-as-I-say style.*

I realize the comparison, for some, may seem to be like comparing apples to apples, but I assure you, the statement is as different as comparing apples to oranges.

The difference is the stock's perception of our presence, not our perception of the stock.

An example would be a corral full of stock, and you enter on one side, pass through the cattle, and exit on the other side. How you present yourself in that crossing will reflect on your routinely practiced engagement with your livestock.

These are examples rather than a test, but I believe it is a way to evaluate oneself awareness and skills.

Do you just get to the other side, in as short of path as possible, or are you thinking about the animal's response, and you choose an altered route instead of the most direct line?

Do you ever take the time out of your busy schedule to practice and influence your stock's responses?

When you're riding through a pasture checking cattle and there is some bedded down stock, do you move those animals because they're lying in the line that you're proceeding on, or do you go around? Or does the stock continue to lay there as you ride through them, as they just chew their cud and watch contently as you pass through?

When you're gathering your stock and you encounter a stoppage, do you push harder and yell louder?

In gathering and moving your cattle, is your approach a

straight line at the stock, or a crossing zig-zag pattern?

Does your stock begin movement when they become aware of your presence?

When you're moving your cattle through a gate, do you position your pickup truck as a wing?

When working horseback, do you engage your horse in a run?

Do your corrals work efficiently?

All of these questions and scenarios are to understand our skills and our stock's preparedness to respond defensively in the midst of a predator's presence.

All of the responses are a reflection of our Stockmanship skills and how we interact with our stock.

These questions are only a measuring stick to make yourself aware of Stockmanship efforts that can affect the cattle's response. The stock's responses can either promote or deny, a *chase sequence as well as individualization* by predators.

When we realize that our actions establish responses, we can better understand our relationship with all of the animals' interactions, even the predators.

This self-awareness relates to many interactions not only with animals but with people as well.

Our conversations and our engagements with people are often closely related to our actions and experiences with our stock.

The challenge becomes recognizing small, but subtle responses to our actions.

Qualifying the stock's response

Often, you hear comments, such as "Boy, those cows are sure wild." Sometimes, there are genetic influences that affect a cow's demeanor. But many times, our actions influence their responses.

Breed influences may present different responses. I would like to point out some breed characteristics.

If you're not aware, the first cow I owned was a registered Brahman. I worked for a fella who had a commercial herd of Angus as well as a registered herd of Brahman. The Brahman breed was unique to me in my younger years, and they were just gaining their brief popularity in California.

The breed has a well-established perception of just being a wild critter and simply being key in an event at a rodeo.

But in actuality, the American Brahman is extremely gentle. The human interactions that they're presented with will dictate their response. Those instilled responses can be set for life.

I have had horses and dogs from early on. I never thought of gathering a tuff range with anything else.

Early one morning, the boss pointed to a pick-up and trailer and said, “Follow me.”

We left the ranch, heading for his range. In his trailer was his herd sire, a very big Brahman bull. My trailer was empty, as I followed, I was just wondering what we were doing.

We arrived at the rangeland corrals. We parked next to them, not backed up to the corrals. Deepening my curiosity about what was on the agenda.

The boss got out of his pickup, reached into the back of the pickup, pulled out a bucket of Cotton seed, set it on the ground, and said, “Hand this to me when I ask.” He then reached back into the pickup and came up with a willow branch, whittled to about two feet.

He then proceeded to his trailer, opened the gate, and called his bull out.

The bull stepped out and looked around, and the bull allowed the boss to lead him away from the trailer, simply by holding the bull’s dewlap. The boss then taps the bull with the willow switch on the bull’s knee. The bull cushed as if it were a camel. He stepped aboard, tapped the bull, and the bull rose.

He looked at me with an extremely large grin and said,

"Hand me the bucket." His grin must have been from my look, I imagine my mouth was wide open, along with my eyes.

I quickly got him the bucket; he tapped the bull and the bull started walking at a quick pace.

I trotted alongside him and asked, what are we doing? He replied, "We're gathering".

"Stay here, when you see me coming back, open the corrals and slap the bucket, there's a bucket in the pickup." As he rode his bull up into the hills, I returned and prepared the corrals, got the bucket and sat and waited.

After a few hours, here he came, riding his bull and leading a herd of cows down a draw.

He rode into the corrals, he slid off the bull and closed the gate after ten or so cows followed him in. Leaving the bulk of the herd outside the corrals staring in.

My trailer fit six cows, and the bosses with his herd sire fit two. We headed out to his home ranch.

I mentioned that I did have a horse, his response was simply, "I know."

He helped me to understand, that when we are working with animals, we are on their timeline, not ours.

At that time, I didn't realize or understand the draw or its connection to the eye. But witnessing it, set into motion an *awareness* that my Stockmanship skills had plenty of room for improvement.

The Eye

We often hear about Horsemanship and even Stockmanship, which refers to the animal's eye and the *draw* of that eye. That draw or that connection is a basic response from the animal.

What the response is, could be viewed as a question from the animal to us as "What"? That connection of the eye can be built on, or quickly lost.

The eye can be either reactionary or can be developed into a draw.

When we speak about the *draw*, our thoughts may go to a lead, the draw may be developed into that, but the draw should be thought of as a willingness or a statement of *"show me."* That recognition is of your engagement.

By recognizing this moment of the animal's question of *"What?"* we as Stockman, must respond to their question with an action of understanding, which is a release, or a pause of that pressure in that moment. After a brief pause, we can progress with our interactions and the objective.

We humans imagine a pause as maybe reclining and having a refreshing drink, eventually returning to our effort. Actually, the pause that I'm talking about is a brief interaction with the animal, with cattle, it may be simply a turn of the stock's head, by pausing and allowing an immediate relief of pressure, it's much like taking a step backwards. That brief pause will allow the stockman to then re-assure the stock of the intended action.

A natural behavioral characteristic of man would be to elevate the pressure at that moment. Equating the connection of the eye as the prelude of resistance. Quickly equating the eye as a *pause and tensity* event.

Building on and understanding the eye is a completely different interaction than the response of *pause and tensity.* Both involve the pause, but understanding that pause, is *reading* the stock.

An example would be developing the *lead.* Let's say that you're moving some stock, and you want to turn the stock. You could turn them by *pressure point pressure*. By positioning yourself on the right side, you would encourage the stock to move to the left. This is *pressure point pressure. Encouraging the stock to move away from pressure.*

Although we interpret that *pressure* point as simply guiding our stock, it's still engaging movement away from pressure. The concept of not promoting a *chase sequence* maybe a small insertion, but nevertheless, it's still movement away from pressure.

Many accomplished stockmen use Stockdogs to manage the stocks movement. Stockdogs are pressure *point pressure*, moving the stock away from their positioning. I often here, I already use dogs. I understand, and I certainly encourage their continued use. But pressure point pressure is only an action that promotes cattle moving away from pressure. Beginning and aiding predators in engaging the chase sequence.

Developing the draw

If you were developing the draw, you would proceed forward at a trot, on the herd's left side.

Increasing your distance from the stock at an angle, while proceeding forward and steadily increasing your distance and separation. *Moving forward, outside of the stock's flight zone. A run would simply draw on a response of fear: Do Not Run.*

While you're developing the *draw,* another helpful suggestion that will be discussed soon is a song, or a statement, or some kind of verbal connection that's attributed to you. Beginning an imprinted response that your stock will set to memory and will have a similar response as saying *"sit"* to your dog.

By the stock's *awareness* of your disengagement, you are releasing the applied pressure, the *release.*

(*An important point of attention at this time would be monitoring your stock's response. If the stock drift away to the right, you are too close. You are within the stock's flight zone. Move further to the left. If you lose the herd posture, return and re-settle the herd group, then re-engage movement and begin again. A natural reaction would be to say, "Well, that didn't work" and simply quit. (The draw is a useful tool of a work with effort.*)

Often, when we're beginning, if we are unable to *'find it,'* we revert to *'forcing it.'* That puts us back to that Do as I say command, *completely ignoring* the eye.

Finding the edge of the flight zone will decrease the stock's perception of applied pressure, slowing the herd group and *drawing* their eye.

By moving ahead of the group, your position would be in that eye forward position, which stops movement, engaging their eye. Allowing a pause in your position, the dwell, which connects you to the stock.

The difference in developing the *draw* versus *pressure point pressure* is more than simply an engagement of moving stock. It's about *actions and reactions, finding rather than forcing. Promoting a herd posture rather than simply engaging a chase sequence in a preferred direction, as well as developing and promoting the stock's Awareness of their surroundings.*

By pausing the group and then turning and walking away from the stock, you would draw the stock's eye towards you.

Turning the stock to the left and developing the lead.

In the beginning, the stocks turn to the left may simply be a turn of the head. But just as all actions begin with a small response, that effort, although small, begins a beneficial *work with relationship.*

This draw is a practiced action, as with any action, it's a response that's a foundation of future responses. The more you practice, the better it gets.

Then, by reengaging the reverse parallel on the left side, as you're facing that group, the stock would reengage movement. You would then simply turn and follow again.

Completing the left turn with the *lead.*

By recognizing that question from the stock, what that establishes is a subtle understanding of a *work with* effort.

Gate Management

This method of the *lead* is a useful tool in *gate management.* When you have reached the position of the stocks pause. You would be able to shorten *the dwell* and proceed to open the gate, leaving the group just standing.

Engaging movement would again be done with the reverse parallel, but it would be done with a smaller group.

By engaging movement with a *pinch-point* separation of a *small lead group and then following them through the gate.*

Once you're through the gate, you would then move to that eye forward pressure point, stopping their movement. Then with lateral movement, you would re-settle that lead group. Providing a destination for the stock that will follow.

As you work on developing this lead, you'll quickly find most often the group that was left will simply follow. If your stock dog is with you, its natural tendency will be to bring that left-behind group of cattle to you.

At the beginning of training, that placement will draw the remaining cattle. Your effort would be simply engaging movement.

I myself prefer stock moving through a gate at a slow and steady pace, rather than any crowding or pushing. *The eye's forward pressure point* enables that control.

Depending on the *stock's sensitivity* will dictate your movement from the lead, back to the following position.

Heavy handling stock will be slower in their actions, whereas sensitive or flighty cattle will react much quicker.

Heavy handling stock will require *more follow, whereas sensitive stock, will require more slowing from the eyes forward pressure point.*

Once you have engaged movement, you'll want to get ahead of the stock again.

By us positioning the previous stock we have provided a destination or a draw for our remaining cattle. By passing outside of the flight zone, we can return to the gate.

By staying far enough away so as not to hinder their approach, we can slowly proceed into our position of metering their speed.

A Faucet

Your position of being in front of the stock, which are proceeding to and through the gate, could be equated to a *faucet*, simply putting you in a position of metering the stock's flow.

Your stance of facing the oncoming stock, puts you in a position to manage their speed. This lead position is historically the Bosses position; it would enable the Boss to get an accurate head count when moving from pasture to pasture.

Your movement a horseback is a simple, slow, and easy sidestep. Simply moving your horse's front feet while holding your horse's hindquarters.

Think of your horse as being a gate, its tail is anchored to an imaginary post. The gate swings in either direction, opening or closing.

This movement would be encouraged with side pressure as well as directing your mount's eye.

Your distance from the physical gate will vary on the stock's sensitivity. Their sensitivity will determine the correct distance.

Reading the stock's response of their eye, will relay if an adjustment is needed in your position.

If a pause is created, back away, if their speed is too fast, move forward closer to the physical gate.

By stepping into the stock's eye, you become a *faucet,* or that closing gate and are simply slowing the stock's flow.

By opening the *faucet* or that gate, you will increase the flow or increase the speed of their movement. Which is simply opening the distance away from the stock's eye.

If you have ventured away from the gate to begin movement with the *reverse parallel*, you can get ahead of the stock by simply repeating *your release* and moving outside of the *flight zone* and passing the stock.

When you're through the gate, move towards that *eye-forward position again, stopping the group. By settling the herd group with lateral movement outside of their flight zone,* you reestablish *a calm herd posture*.

This will allow you time to close the gate and resume and redirect movement to your intended destination.

A mentor

One of my many mentors used the *draw of the eye and the lead* quite often. He also used it with his personal call. His *call* was a bit like a song, it kinda went like, Aaaa....*Saaayyy....Yoooo...*

When the stock turned towards him, he would sprinkle a handful of corse ground salt on the ground as a treat for his cows. Strengthening the cow's draw with a reward.

This Rancher worked on many tough and expansive ranges. It seemed as if he always had a pocket of corse ground salt. If he ran across a noxious weed on his range, he would sprinkle that undesirable plant. Mitigating the risk of an invasive species taking hold.

A pocket of salt served in the settling of his stock and strengthening that *work with relationship*, as well as benefiting the quality of the range.

Building a *working relationship* with his stock as well as maintaining and improving his environment.

All related to that *draw of the eye*.

A colt

It's much like when you're starting that young horse. At the beginning of groundwork, you have successfully engaged movement and then you pause because you're looking for the youngster's response.

Often, we analyze our effort, such as engaging the Colt's movement by a timeline of what we believe is far enough for our satisfaction. Rather than realizing that the engagement of movement itself, was the success.

Once the colt has engaged movement, our next objective needs to be an acknowledgement and how quickly the colt yields its eye.

What I want is a tip of the ear, a small look towards me, or licking and chewing.

Then, the animal's response will be a connection to wait and see what your request is.

For me, with a young horse, the distance of the engagement of movement is not the point of the effort. The point of the effort is to begin that *work* with relationship. More than the movement, I'm looking for the conversation.

As to that youngster whom we have just engaged movement with, when it yields its eye, I would extend my lead arm and give a cluck, and with my other arm and flag, I would pressure it to continue the engagement of movement.

(For myself, I don't like quitting on the first yield of the eye, although recognition at this time is important. My experiences have

suggested that an immediate release creates and develops a quitter when the going gets difficult. Just a note: I have also found that response to cross species lines.)

The next request, or the next yield of the eye, I would ask for a slowing and continue the request until the stop happens.

I would do this by recognizing the eye and stepping away from it, backing up, drawing the youngster towards me, the draw.

(Just a brief mention about movement on young colts, Fear is something we should overcome, not install. Flight is a response brought forward through fear, an instinctual response. When I begin these youngsters, I want to develop a working relationship. So, we'll get to a lope, but movement in the beginning is simply a walk or trot.)

As the youngster stands, allow time for the dwell. Often, in the beginning with this youngster, I will walk to it and give it some contact praise. In that time, I may reach across its neck and grab the tail of the lead and with my other hand place the rope across the youngster's nose.

By pressuring an even pull of both hands, I will pressure the youngster's nose with a light pressure. The slightest tip of the nose or flexing of the poll will get an instant release. On that achievement, I will end my efforts. This session has been a successful one.

Done, isn't really all done

Ending the session has its methods as well, it's not simply walking away. I want my horses to want to be caught. My horses don't get grain as part of their daily ration. They get grain when they reach the tack room after being caught, and they receive grain during unsaddling after the day's effort.

The grain is a tool for establishing a connection. *Rewards and consequences, working with* rather than just existing together.

Before turning my horse loose, I lead the horse through the gate and turn it towards me. I then take a moment to rub and scratch my mount. I'll continue until I receive an acknowledgment, just as we talked about before. I will then try to take off the halter, I want no attempt of movement.

I'll slip the halter off and step away. What I want is the young horse to watch me leave. Not me watching the youngster leave and the youngster being happy for its release.

The release, whether for the young horse, or even in working with our stock, is as important a part of the *work with scenario as any pressure.*

The proper release is about finishing on a positive note. Establishing that the session was a positive encounter. The proper release is as important as the entire session. This release is important for *horses, cattle, or the dogs we work with.*

It's been my experience that the length of the training session isn't important, but rather the frequency of those sessions and quitting on a positive note is the key to developing a working relationship.

Recognizing the response of an eye's reaction is key to understanding the development of a *draw.*

Missing that recognition is an opportunity lost and as time goes on, the animal will simply quit asking questions. Developing a reactionary response rather than a *draw.*

Chapter Five

Reading the Stock

The challenge in identifying the cattle's eye is understanding their postures. This is *reading your stock.* Just as with the young horses, the cattle respond to our engagements and give signs of that acknowledgment.

By recognizing your stocks *suggestions of intent,* you're *reading your stock;* that interaction is the goal of good stockmanship.

Their postures are a telling sign of *suggested intent.* By acknowledging and reading their posture, we can interact with them in a way that guides their responses. Our interactions establish either a *work with effort* or *do as I say response.* In developing *predator awareness, it's better to guide, rather than react.*

The cattle's response will be demeanor, tip of the ear, height and carriage of the head as well as pace. The cattle don't have the licking and chewing as a sign of acceptance as horses do. Their acceptance is a stance, that stance is a wait and see stance, the stocks response will be to your applied directions.

This *stance* is a direct connection to that *eye forward pressure point,* which is the *draw of the eye,* which creates pause and stops movement. The animals focus will be directed at you.

This pause is the same, whether the stock is preparing to challenge you, or work with you. If the stock are known to be sensitive and an individual has been individualized, this is a time for caution. Challenging this individual, such as moving towards it, is the same effort as a matador waiving a cape in a bull fighting ring. Your movement towards this individual is simply challenging it.

This is *pause and tensity,* your actions either elevate or de-escalate the *tensity. (Pause and tensity is the first engagement of the instinct of self-preservation. The pause stops movement and begins the tensity. The tensity is an analyzation on an instinctual level of the perception. An escalation of tension will engage fears action of fight or flight.)*

If your actions have been that *do as I say* type action, then their response will be a reflection of that: fight or flight.

If your actions have been that *work with* interactions, their responses will be your guided directions.

We just discussed *movement being intentional, stopping or pausing is also intentional.* Understanding that our actions have responses, our objective is understanding those actions and applying them in a way that achieves the response we are seeking.

If you have a question of the stock's intent, it's always best to ere on the side of caution.

I realize that with experience, the animal will begin to predict your actions by your behavioral tendencies, but that interaction is the Summit of a *work with* understanding. Just as with any relationship there will be misunderstandings, but those will be far and few in between.

In this advanced stage of *working with,* it's hopeful that if a problem arises, we will recognize that our interactions were simply misinterpreted, rather than reverting to that *do as I say* method.

In this working relationship that we have developed its important to be aware that the animals within this partnership don't ponder malice, but rather simply offer compliance to our requests, just as partners working towards the same goal.

The Call

Several of my mentors taught me the value of verbal communication with animals. I realize there are a few who just started grinning and are about ready to chuckle.

But I'm not talking about human conversations of words, but an animal's conversation of responses.

There are certain things that just create conversation. Take a young mother and her first trip to town as a mother. People's responses are an interesting phenomenon. People who are complete strangers spark up conversations with the new mother, all with a variety of praises.

That new mother phenomenon doesn't require another new mother to engage in such a conversation. It happens from children to the elderly, they're all drawn in.

That *new mother* relationship is amazing.

Now, some of you that are still grinning but have been holding your chuckle, get ready.

That *new mother* relationship, as I just mentioned, exists with animals.

Many are thinking, why of course, that mother is going to have a great interest, it is her calf.

But I wasn't talking about the mother-child relationship in the above, but the draw that the child creates between the mother and others.

That draw of new motherhood is the connection. One may say it's simply curiosity. Or that the strangers are simply looking to hear a new story. How you perceive the occurrence isn't the point, the point is, that the new mother and her child draw attention.

Some of you from the Ranchers side of things are certainly aware of this interaction. Many times, a mother will adopt another mother's calf and treat it as if it's her own. I've heard a few even refer to it as stealing. A lot of the time, it's because she hasn't calved, and her motherly instinct just requires a calf. So, she gets one.

It's not really the action we're talking about, but it's connected.

That draw, whether to the calf, by a wanting mother, or to the *new mother* and a stranger, is the *call.*

By mimicking a calf verbally, urrrr…, we create an interest not much different than the new mother's first trip to town and the stranger's interactions. That *call* or the engagement, what it will cause, is a *draw of the eye.* That animal, that we are *calling to,* will stop and look.

I am not sure if this connection is instinct or curiosity, I only know it creates a *draw of the eye.*

That stop, was from a *pressure point pressure* directed through the animal's ears to its *eyes forward pressure point. Stopping movement and creating the draw of the eye.*

With the ability to engage an animal's *eye,* by the use of the *call,* we pause movement. Why and how we use that pause to our benefit is our choice.

The functionality of the Call

One of the larger ranches that I worked for, brought Mexican steers in for the winter months, grazing on California's winter green ranges.

These steers weighed around four hundred and fifty pounds. But their frame was the height of a seven-hundred-pound domestic steer. Granted most of their weight was frame with a lot of leg.

They definitely could run, their abilities to contort was a challenge in tying them down, better suited for youth, rather than experience.

Keeping these steers healthy and gaining weight, is what I was hired for. Of course, this was in my younger years and getting western is a part of youth, so there was some need for speed.

But I also worked with seniors, like myself today, they appreciated my ability to throw that long shot, completing the objective quietly and safely as a team rather than an individual effort.

In order to accomplish these long shots, I prefer the animal to be standing or walking. My point is, the *call,* which *draws they're eye, creates opportunity with such a pause.*

This *call* isn't a correlation to motherhood, but rather to the *draw of the eye.* When I used this *call on Mexican steers* it attested to its functionality as a useful *Stockmanship skill.*

This method crosses species lines, although we're mimicking a calf, every ungulate that I've tried it on, at the very least had a response, it created a pause.

One cow elk actually presented herself in an aggressive manner.

It Even Works On Turkeys

A cow boss and I were in the back country sorting out pairs from calvy cows.

We had gathered the field into a corner and had settled the mixed group. So, each of us would pull pairs and take them to another location that was a distance from the calvy cows. The intent was to move the pairs to an adjoining field, as the gate was only about a half mile away.

The reason for this effort is what we just talked about: the cows would adopt or steel another's calf. By separating the pairs from the expecting mothers, we kept the mother's interactions straight.

We used the call to get the proper mother and calf to pair up.

While we were both busy focusing on the task at hand, apparently, our calls created a curiosity for a group of turkeys. The gaggle had wandered in, and neither the boss nor I noticed. I don't recall which one of us made the call, but one of us did.

The day was an opportunity to put quiet time on a colt, just plenty of quiet miles.

So, both the boss and I were riding a young colt.

Well, the group of turkeys heard our interpretation of a calf, and in unison, the Toms within the gaggle let loose with an almost deafening gobble. Needless to say, our quiet day quickly went western. The cows, being accustomed to a turkey's presence, simply appreciated the show the turkeys presented to them.

After a few minutes we both regained our composure and settled our mounts and were back to sorting pairs. We were both joking and laughing about the turkeys sneaking up on us. We couldn't believe we didn't notice them. So, we were back to calling to connect the mothers to their calves. The turkeys would gobble, but our colts as well as ourselves, were aware of their presence.

We were both commenting about the turkeys, they were now in amongst the cows, just walking through them. We would make our calf call, but the turkeys wouldn't respond.

We were closing in on our objective and were close to the final few drafts. Close to completing the sorting. The boss was taking some time out to acquaint his colt to the turkeys. He was driving a herd group of about thirty turkeys rather nicely.

I thought at the time it would be a good effort to move my colt closer. Getting him accustomed to the gaggle. So, I did.

There, the boss and I set, looking at our herd group of turkeys and having a laugh. We were talking about the turkey's response to the calf call. Although we couldn't get them to gobble, It still had us chuckling.

Apparently, the boss wanted them to gobble, so in his best imitation, he gobbled.

Neither the colts, who were standing quietly, nor ourselves anticipated the turkey's response. The Toms instantly spun towards us and plumed their tail and wing feathers, making quite a rattle.

Our quiet went to full-blown rodeo instantly.

I noticed that the bosses horse went for the corner that we were sorting out of, moving the bred cows down the fence towards the previously sorted pairs. My horse was heading for open country, after looping him back, I realized the bred cows reunited with the pairs. The boss and I soon met up, it seems that neither of us found the turkeys to be so humerus anymore.

The day was shot; this story is the day's only achievement.

The turkey story relates that *the call* may not always work, but it's an option.

Binding pairs

Another tool that has a lot of opposition for many in sorting pairs is a dog.

The *predator-awareness dog* has attributes that fit well into *binding pairs. Sorting pairs* is a routine practice on many ranches. Many ranches run steer pairs separate from the heifer pairs. Many believe that the steers gain better and the heifers can learn the country they may spend their life in. Learning such things as water location, trails, salt and mineral placements.

Binding pairs is a bit different than *sorting pairs*. It's about strengthening the mother's awareness of plausible threats and elevating her attention to her calf's well-being.

The effort is a piece of *predator awareness training.*

When you're presenting the *pseudo predator* amongst new pairs, the action you're looking for is rather than the mother focusing on the dog, we are encouraging her to seek her calf and *bind to it.*

The handlers control as well as the dog's distance from the edge of the flight zone, as well as the dogs speed, urge the mothers in uniting to their calves. A slow and stealthy type engagement, as from a quality stockdog only aggravates the mothers, often resulting in the mother chasing the dog and stirring the other cattle.

The difference is the dog's movement. A quick-moving dog outside of the flight zone engages the motherly reaction of seeking her calf's location. Whereas the stealthy actions of a well-mannered stockdog are perceived as stalking and require a defensive response from the mother, which is the action of chasing the perceived threat.

This engagement is an opportunity for a training session on *predator awareness.* When we're establishing the defensive posture, one of the first responses that we are trying to achieve is the mother seeking her calf and engaging in the *Standing Solution.*

The awareness from the mother is from such presence that may threaten her calf, which is an instinctual response. Strengthening that faded instinctual reaction.

With speed and distance, these sessions encourage the mother *to bind* to the calf rather than engaging the predators.

This description of interactions would be to cows that have a *dog awareness* already. If the stock are not dog broke, or have been previously predatorized, the *predator awareness* assistant won't help, it will only promote unsettledness. Preparing the stock for working with dogs is better done with bred cows, but if you're like me, I'm always behind. Settling pairs, that aren't accustomed to dogs, or even cows that have been predatorized is just adding another step into the process. That step is using the technique of *predatory desensitization and* developing *the herd posture.*

In the effort of settling previously predatorized stock, the preferred introduction is best done with a fast-moving dog whose control is good. The effort is about desensitization.

Predators will capitalize on a non-beneficial response of the mother abandoning her calf while she challenges the predator. Leaving the calf vulnerable and unattended.

By addressing and developing this *binding,* we circumvent a predatory opportunity.

A single dog fits this session; I prefer a *soft dog* in this situation. When I say *soft*, the dog should be very aware of the cattle's flight zone and should be easily controlled. If the cow moves towards the *pseudo-predator, the pseudo-predator* should retreat. If the cow continues the chase, the *pseudo-predator* should create a fast lateral zig-zag pattern of pressure to encourage the mother to return to her calf.

(This action is not "chasing cattle." Rather, it is binding a mother cow to her calf.)

This objective of binding pairs is better done with one rider and one dog.

Reading the cows is key to *positioning the dog properly.*

To many human interactions, such as multiple riders, have the cow's attention wandering from rider to rider, then back to the dog, only escalating her tensity and not focusing specifically on her calf.

What happens is the tensity just elevates into an instinctual response of *fear.*

We want to present a calm, safe alternative to that *instinctual reaction.*

Encouraging a calmer response of seeking and *binding* to her calf. Promoting a *solution to a predator's presence.*

I realize that the *natural obstruction* in many of these *skills*, is working alone. Many are comfortable with others; I realize that having companionship is important and being alone only promotes un-assuredness. But, conversations and even multiple presence just require the stocks attention to wander, rather than focusing on the task at hand.

That cow's focus is the task; her focus should be *binding* to *her calf.* The key is the dog's speed outside of the flight zone and the mother's response. By the reduced presence, you keep the interaction between the cow and the *pseudo-predator.*

The dog does not engage the mother but rather avoids her, offering to the mother a concern with its presence. The cow will soon realize that her calf is unattended and return to it. You, as the trainer, can encourage her re-uniting to her calf with the use of the *call.*

When she reaches her calf after this session, all of the pressures need to stop, allowing her time to become comfortable by her calf, *promoting the standing solution.*

Further enhancements of *binding* will be built on later with the training of the herd group. This response is a *building block* of the *defensive posture of the herd group. Not leaving her calf unattended. As well as seeking her calf on her awareness of presence.*

Canine introductions

Oftentimes, we're calving in the winter months, and the pregnant cows are being fed hay on a daily basis. This time of mixed pairs and calvy cows is not the time to be introducing cows to canine presence.

The best time to begin such an introduction is after weaning her previous calf. At this time, your introduction is to a non-mother; there is no calf for her to be concerned about. Her bred status aids us, as her actions are less flighty. The introduction can be completed by the development of the *herd group,* but if the cows are accustomed to domestic canines, this winter or this spring calving ground, where you're feeding on a daily basis, is a good location and opportunity to encourage the *standing solution as well as binding pairs.*

The objective is that when the mother sees a questionable situation, she *seeks her calf and binds to it, then proceeds to the defensive posture of the herd group.* In the first sessions, that simple recognition of *presence* and then looking towards her calf requires a release of pressure. Ride away, watch from a distance. Did she go to her calf? If she did, you're done for the day. If she didn't, move in again and repeat. If, by chance, she still avoids her calf, you are better off leaving further training for another day.

Future pressures and the *pseudo predator's* presence should achieve the result of the mother looking towards her calf and *binding to it.*

As with any session with any species, developing a response is a progression of small responses that are built on one another to accomplish the desired results. This fundamental action is intended to establish an awareness of a predator's presence. That awareness is to instill a response of seeking her defenseless calf and not engaging the predators.

The calf, when left alone, is simply an opportunity for predators. When multiple predators are present, the individual calf is quickly lost.

The task is about strengthening that bond of motherhood, not about creating a stockdog or controlling movement.

I occasionally hear comments about a cow being equated to a good mother because she'll get you if you venture close to her calf. A cow's aggressive nature is often not a beneficial quality if she abandons her calf to chase a predator, that defeats the binding effort.

Only leaving the calf vulnerable and unattended. The *binding method* is a correction *to this overly protective mother* and her action of engaging the predators. The dog's *speed, as well as distance,* encourages desensitization, which removes movement and presents a reaction to a predator's presence of *binding to and standing by her calf.*

The Lead mare and Sargent cows

When we are dealing with multiple animals, there will be a lead within that herd group. Most often with horses, it's the lead mare.

Most often with cattle, it's what I call the *Sargent cow*; she is the cow that others are drawn to. The anchor to that *herd group.* The lead cow, so to speak. Although her actions have much more interactions with the herd than just it's movement.

A *sargent cow,* in comparison to a horse, is in the same position as a *lead mare.*

The *Sargent cow* is the *anchor* of the *herd posture.*

When I began *predator awareness,* I soon became aware of this very interesting occurrence. The first signs seemed to be that certain individual cows seemed to be less reactionary to the *pseudo predators'* pressures. Their response was not the response of fight and quite the opposite of flight.

This posture seemed to me to be a defiant stance directed toward the *pseudo-predators.*

They were routinely the center of the *defensive posture of the herd group;* one might even say they were the anchor to that *stance.*

When they were captured on a trail camera, they were often the *lead cow* in a group that was going to water.

Another noticeable characteristic was nurseries. Young calves were left in a group while their mothers went off to graze. A mother and sometimes two would remain to watch over the defenseless calves; these cows were often *sargent cows.* Watching over the calves until the other mothers returned.

Another interesting action was when the *defensive posture of the herd group* was engaged; the *sargent cow* would continually watch the *pseudo predators.* She would posture as if she were going to chase and fight the *pseudo-predator.* But in action, she only blocked a connection between the *pseudo-predators* and the *herd group.* If she moved towards the dogs, it was more of a slow stalking pace rather than a *chase. Posturing but never abandoning the defensive group.*

The *pseudo predators,* though, are encouraged to continually move, so the *sargent cow's* stance just resembles a defensive line. Blocking further encroachment by the *pseudo-predators.*

I'm not sure if one could choose and develop a young filly to become a *lead mare.* But, through the development of *Predator Awareness,* I have witnessed the *pseudo predators'* efforts and witnessed the cattle developing this skill of the *sargent cow.*

A unique occurrence

Our ranch's location requires winter feeding, so my neighbor and I have a common fence line. One morning, he called me and said one of my cows was in his corrals, waiting for me to pick her up. I told him I'd get her as soon as I finished feeding. His response was, "ok, she's an easy-going cow; she has a mellow temperament."

When I made it to the neighbor's corral, there she stood. I recognized her quickly. Her actions had me questioning what was going on. She's about a month away from calving, but she was repeatedly calling her calf. We have a significant amount of foothill abortion in this area, and as she's a spring calver, foothill was certainly a concern, but she showed no signs of aborting.

I didn't back up to a loading alley but simply backed up to a gate into this large pen. I swung the gate open, walked out into the pen with a young pup, and the cow jumped into the trailer. She almost seemed nervous, and she continued her calling.

I decided that I would leave her in the trailer when I got back home and take a trot around the fence line to see where she escaped or find what was creating her uneasiness.

So, there is a ridge that runs through the ranch; the cows have access to this area, and it's adjacent to the Klamath Basin Wildlife Refuge. I hadn't ridden for a bit, and I thought that I'd just check out the perimeter fence. We had around a half inch of snow through the night, so seeing where she escaped should be easy. After checking about a mile of fence, I saw her tracks quite always in front of me, not at all where I was thinking I may find a hole. The tracks came from my feed ground; at that time, she was heading east, the opposite direction of my neighbors.

When I made it to the fence, it was stretched out but in-tacked. Her tracks told the story of what happened. She came off the feed grounds apparently at a trot, found a weak spot, forced her way through the fence and onto a county road, and then headed west. Heading in the opposite direction that she left the feed grounds in, and then turning towards the neighbors.

I tightened and repaired the fence, all the while thinking how strange this entire mourning had been.

So, I stepped up and headed to the gate that would allow me and my mount to get onto the county road and track her, if for no other reason than simply curiosity. I was surprised her line never wandered; even though there was dry feed along the shoulders of the road, she never stopped to browse.

A Bad Sign

I continued tracking her, and then, about a mile ahead, I saw birds. Birds aren't a good sign. As I got closer, I definitely knew there was a problem. The birds were ravens and magpies. Those birds told me that there was definitely something dead. I thought that somebody's dog died, and they dumped it in the country or something along those lines.

When I got to the sight, I realized it was a depredation. The predator had run this three-hundred-pound calf through the fence and into a ditch that ran alongside the county road.

The calf belonged to the neighbor, the same neighbor who called me to begin this episode. He lives a way from this feed ground, so I called him and told him the story. He couldn't believe he missed the missing calf.

Around this calf were my cow's tracks; she was definitely chasing something around the calf carcass. When I back tracked, I quickly realized that it was a wolf depredation.

My thoughts quickly went back to my cow; this morning was unique, to say the least. But her actions attested to *predator awareness,* and it strengthened my thoughts of *sargent cows* and their frontline tactics.

That cow saw the depredation happening, and she went over a hill and through a fence and made an effort to defend the calf.

I'm not saying that her efforts were admirable, in fact, if there was a pack, I would have probably lost a cow as well.

Her actions were definitely a response that's unique.

What Cow?

The neighbor didn't know who the mother was and asked me to keep an eye open. I told him I would try and figure out what was going on as neither of us saw a tight bag or a mother looking for a missing calf.

After some time of counting wet cows versus nursing calves, there was definitely one calf missing. So, after a while, I found the mother who was letting any calf nurse. I told the owner about her and mentioned at least she'll be easy to graft. He thought that the sale was a better spot for such a mother. The day's events were a unique occurrence. I lost that *Sargent cow* the following year to bloat. She never repeated her individualization, but my deterring efforts expanded to cover my neighbor's ranch as well as my own. The posture of the *sargent cow* is still seen when you are training stock. But often, those postures are subtle. I have found that the introductory period is often the easiest time to define the lead within the group. With further training, most of the cows establish this behavior and become themselves *sargent cows* and become comfortable with the *standing solution.*

Our efforts in establishing a *draw* will be recognizing the *sergeant cow* and distinguishing it from another cow's reactionary response.

The *sergeant cow* will not be the cow who initially engages in flight. But she may be the cow that initiates a fight sequence. If she pursues the dog, it won't look like an all-out chase but rather a slower movement in the pseudo-predator's direction.

Most of the time, the *sergeant cow's* action in a new situation will be a stance with no movement, an anchor for that herd group.

Recognizing the cow who is posturing for flight is important in training a response. That individual is the one that will present a trainable moment. Although we are working with that one individual, that effort will be an example for the others within that *herd group* that individualization is not a good thing.

The Difference Between Dogs

There's a difference in canine behavior, and that difference is about the dog's actions. A breed is often associated with specific performance. Queensland healers are generally heeling dogs; they address the rear of their intended target. Whereas heading dogs address the front, or the head of their target. Breeds like the Black Mouth Cur will perform differently; their action could be viewed as a baying action. Grouping their target into a group.

Probably the dog that's most admirable is the Border Collie. It's stealth and herding abilities are simply a joy to watch. The Border Collie is simply an extension of the stockman.

Their fault is they handle cattle from a pressure point pressure. Let's say you're moving some stock, and you want to make a left turn. You could send your dog along the right side of the cattle. With that movement, the stock would recognize that pressure and move to the left.

The stealth that the Border Collie presented in this action, which generated movement, is admirable. But admirable qualities don't promote the *Standing Solution,* no matter how pretty they are to watch. That stealthy action that we all admire is simply perceived by predatorized stock as a stalking action. That stealthy action of stalking then generates fear within the stock. The quality stock dog only promotes a *chase sequence* in our preferred direction.

In order for the action to create the *Standing Solution* we need to pressure the stock into a group. In order to accomplish this, we need a dog whose objective is just that: removing movement. Much like the Black Mouth Curs action. Quite different than the quality *stock dog.*

The two objectives are different. One promotes movement; the other eliminates it. The *predators* require an individual, which is the first step in creating the *chase sequence.*

I often hear, I already use dogs, I understand, but the two actions and the stock's reactions are quite different.

Then the response is, what am I supposed to do? Stop using my *Border Collie?* Hopefully, you'll realize how the two actions are different and adapt your skills to not *promote a chase sequence* that benefits the predators.

The two actions, *driving or stopping,* are completely counterproductive to each other. The *quality stock dog* and the *predator awareness dog* shouldn't be used together. Their actions and effects with the stock, when used in combination, only create confusion.

The common response is, if I create the *Standing Solution,* how do I and my *Border Collie* move the stock? By creating movement from the lead and creating a follow with the *reverse parallel.* When you lead the cattle instead of driving them, a *natural characteristic of Border Collies is to bring the stock to you.* You lead they follow. Not promoting *pressure point pressure,* but rather moving a herd group rather than keeping *individuals grouped together.*

Helping Conception Rates

Bulls are for most Ranchers, an expense. By selecting sires with desirable genetics, the Rancher increases the value of his or her marketable calves while increasing the performance of the retained stock. These improvements have an associated cost; the better the bulls with better EPD's (estimated production data), the higher the price.

Often, these replacements are lost to a problematic behavior of bulls fighting or another bull hitting a mounting bull. Injuring the individual bull and making him ineffective for the breeding season or sometimes never.

I'm not sure whether it's a trained response or a trait simply bred into the predator-awareness dog. But my dogs have a true dislike for cattle, whether cows, bulls, steers just butting heads or anything that resembles fighting. Their response is an immediate effort to break up the altercation. Followed by a direct action of encouraging separation.

Often, another behavior of bulls is to bunch up with other bulls. In baseball, it's called the Bull Pen, where all the pitchers warm up until called into the game. Another interaction is the local watering hole (a bar), where all the men gather to solve the world's problems.

This action of bulls grouping is a concern. This grouping leaves the cows unattended. By separating individual bulls and placing them with groups of cows, you're simply encouraging them to do the job that they're intended to do.

The *predator awareness assistant* seems to have a built-in understanding that bulls have a job.

Possibly adding another economic value to the effort of deterring predators.

Chapter Six

Show Me

Animals are reactionary; their responses are from a *show-me* interaction. A simple explanation of the *show-me interaction* would be training a young dog to sit. Although you say "sit," the pup has no clue about your request.

By physically engaging the pup *(showing it)* and reinforcing it with verbal commands, the pup soon recognizes your request.

As to our stock, that *show me* interaction can be a good thing or bad. We, as the *stockmen,* are the ones who should be establishing it first. It is with our actions that we re-engage a preferred instinctual response.

Predators can also set up a response in our stock; that reaction will be *fear*. By allowing the predators to establish *fear* first, the cattle are never presented with a release or a *solution* to such pressures, only the response of the *self-preservation instinct* and its response of *fear*.

With *predator awareness training* and pressuring this individualized cow and releasing that pressure on her proper response, she finds relief within the herd group. With that action, we establish a response and relief to such pressure. By showing an individual stock where to find such relief, we provide a solution. By presenting our *pseudo-predators* to the individualized cow in multiple sessions, we encourage her to seek relief before the *pseudo-predator's* actual engagement, instilling an awareness of their surroundings.

The reasoning for the multiple sessions is that the stock seeks the *defensive posture* when they become aware of a predator's presence. I call this *herd awareness.*

Although we have been dealing with one individual cow at this time, we are actually addressing the entire herd group. The interaction between us and this individual cow is a *show-me training session* for the others as well.

The Pseudo Predators' severity

Sometimes, when I'm discussing this show-me interaction, someone will begin talking about how good a dog is by the severity of its contact, it's bite.

I don't like biting; it defeats the purpose. Biting establishes *fear. Our efforts are about replacing the response of fear and replacing it with the calm resolve of the herd group and the standing solution.*

The quality that is most important is speed and control. The stopping ability of the dog is the blocking of movement rather than biting. Occasionally, biting happens; for myself, I discourage that action. I don't encourage *fear but rather promote an alternative solution, which is the comfort of being within the herd group.*

When we begin shedding individuals, often the perception is the dog is just chasing the individualized stock. But in actuality, the *pseudo-predator* is creating *fear* and showing the stock that relief to such a *fearful* situation is joining the *defensive posture of the herd group.*

The response we are seeking is that the *herd* equals a *calm and safe* place for relief from pressure. A trainable moment, as well as a *trained directive action.* That presentation of the *herd group challenges the wolf's self-preservation instinct.*

Seldom does a wolf assault its prey victim head-on. The *chase sequence* puts the *wolf* in a position of control. The assault of predators most often occurs from the hind end of the victim.

Game officials often skin deprivations hind quarters, searching for bruising or damage created by the predators in promoting the *chase sequence.* I often ask game officials how many times they have seen wolf conflicts to the head of deprivation. The common response is "not many." With further questioning, it goes from "not many to never."

A small but significant fact is that the *standing solution is an effective response in mitigating confrontation. Cows facing a predator provide a connection to the predator's eye, which causes an instinctual response of a pause and tensity event within the predator and* is an effective deterrent.

Moving past the response of fear

There will be another cow's reactionary response in the beginning, because we have not established the *Defensive Posture of the Herd Group* yet.

When we move ahead with further training, we will be required to engage in individualization. This is shedding. The predators will also use this technique in searching for the weakest link.

So, individualization will happen, but that's the reason for these efforts. Avoiding future Individualization by a predator and strengthening the *Standing Solution.*

We have just chatted about the animals responding to that *show-me* interaction. That's a basic action that solidifies a response. The response will be to future engagements to draw on *what they've been shown.*

That *show me* interaction can also be a response that's not beneficial. If the predators set the *show-me* response, it *benefits the predators.* But if our actions promote a non-beneficial response, it's easy to understand *that those responses also benefit the predators.*

Finding rather than forcing

Our interactions are about cattle and preventing economic losses to predators.

By developing a *response to perceived pressure, we engage both sides on an instinctual level. Finding a comfortable solution to a predatory presence through behavioral responses rather than promoting such a non-beneficial engagement as the chase sequence.*

Our position

We had just discussed the *eye* and its connection to our *Skills* of reading that *eye* and *working with* it in relation to *Stockmanship as well as Horsemanship.* We then discussed the lead. These two words, by their interactions, are our connection and guidance in establishing good *Stockmanship* and *Horsemanship.* When we are developing a *working relationship,* what we are trying to establish by connecting to the individual is a position for ourselves as the *lead* in that conversation.

What we are accomplishing is that our engagement creates an intended response from the cattle. Rather than reacting and responding to their movements, we direct the conversation, and we guide their actions.

The intent is that we, the stockmen, become the *interpreter* between our stock and the predators.

Our goal is to re-establish *instinctual responses and defensive herd postures.* We want to realize that when that animal offers its eye, it's asking the question, what? By recognizing and responding to that question, we are developing a *work with relationship*.

Our response to that question is a pause of our applied pressure. Our intention is to develop a herd action rather than an individualized response. We don't often accomplish this task within the corrals, but we can damage it there. Promoting individualization.

An example would be moving a group of cows within a corral. You're moving these cows to another pen. You open the gate and proceed to the closest cows instead of going all the way to the rear.

By beginning movement from the lead, you have started that conversation of *we are going there.* By then encouraging the others to follow with the reverse parallel, you create a follow, rather than trying to pressure the entire group from rear pressure, pushing them through the gate, *the Chase.*

Creating movement as well as developing the group, rather than simply chasing the herd through to the next pen.

Drawing and developing that connection of the *eye.*

There may be a cow within the group that doesn't want to go through the gate; she stops she looks away from the gate.

By stopping and moving with a disengaging movement, which is a lateral movement of no increased pressure, we block her intention.

By pausing and not applying additional pressure. Her response will be to follow the other stock, denying individualization and promoting the *herd group.* When her knee closest to the gate bends, we may return to the reverse parallel, encouraging the remaining stocks movement.

If we were to yell or push harder, we would be drawing her intentions to the break, promoting individualization. Forcing her response to leave the group rather than follow it.

That action, although unintentionally, promoted individualization; our only intention was to move the stock to another pen. The consequence is that we aided the predators in individualizing the stock.

Undermining our efforts and training of the *defensive posture of the herd group* and promoting a beneficial response for the predators.

By recognizing the intention and addressing it, we become the *lead* within that relationship. Within that relationship, when a question arises, we must show guidance to develop trust.

By *developing trust,* we may have a *calm interaction* rather than a *reactionary* one. We will have *"that conversation"* rather than a shouting match.

Promoting a herd movement rather than an individual action.

"As the lead, we are guiding the stocks responses; we're not reacting to their actions."

The animal kingdom is reactionary; it is not judgmental. In that animal's request for guidance from you, the *lead,* it's irrelevant whether your response is right or wrong. The only thing that matters is recognizing that request and responding. Simply pausing for a brief moment is a recognition of the stock, which will then respond rather than ignore.

Reading your stock equates to reading their intended action. The stocks eye, as well as their posture, lends understanding to those intended actions.

By recognizing *intention* and then pausing and redirecting our pressure, we guide a calm and efficient response.

By encouraging the movement of the herd group even within the corrals, we interject a suggestion that individualization is the wrong action to applied pressure. Promoting a response rather than a reaction.

Effective and controlled interactions are the *work with method.*

A Common Error

Often, a solution to an obstruction of a smooth flow is simply pushing harder, reverting back to "*what's required here is a bigger hammer.*" *Compounding* the stocks response into a *pause and tensity* event rather than a *work with* solution.

Promoting *a reaction rather than a response.*

The pause of movement locks the stocks feet and begins escalating the *tensity.* By continuing to pressure your drawing, the *stocks eye* to that *pressure. If you're at the back, it's drawing that pressure on you.* In most cases, a complete opposite of the desired result.

A Correction

If we understand that applying more pressure in a *pause and tensity* event promotes actions that undermine the desired result, we could then say to ourselves, "Gee, if I applied pressure at the lead, the stock would proceed to the intended destination." Simply flip-flopping the applied pressure and engaging movement in the preferred direction.

That realization is an example rather than a solution. The solution would be to avoid such a reaction and present a *work with event* that engages proactive responses.

First would be the discovery of what created the pause and

addressing it after resolving that issue. Beginning the lead would simply be applying the *reverse parallel. Promoting movement from the front and developing a follow for the remaining stock. Proactively correcting such a pause and tensity event.*

The Response

In the above, we have been discussing the stocks responses to perceptions and the previous experiences that have instilled those reactions.

Be assured that a predator will only begin the interaction with the stock abruptly, with a threatening action, in an attempt to engage an instinctual response of the self-*preservation instinct* and its responses of *fear.*

It is important to establish a response with our stock, prior to the predators.

Our actions shouldn't encourage such flight and individualization, which aid the predators in their objective. Sometimes, our management requires such individualization; it should be an anomaly rather than a routine.

After we have completed the management effort, the day, as well as the stock, are spent. Within the following days, there should be an effort to reestablish a herd group. That would be a tuneup to re-engage and strengthen the *Defensive Posture* and the *Standing*

Solution.

Many times, we must simply turn the stock out to go back to their job of grazing; I realize there are many things that occupy our time. The *tune-up* can be done in big country and worked in with monitoring.

The objective of a tune-up is a brief engagement of pressure and the witnessing of the response. With the proper response, you can simply move on. If, by chance, something such as flight or individualization happens, then you would require more training in the *Standing Solution.*

The stock's responses, are what tune-ups attest to whether the cattle are prepared for predators.

As humans, we have the ability to verbally communicate. The animals converse through sensory perceptions, actions, and posture.

Although the animals don't communicate as we do, I'm pretty sure that the majority of us understand that when a dog is growling and baring its teeth. It's probably not a request to be pet. That's a simple, understandable interaction. I don't speak many languages, but I understand with effort I could develop a better understanding of those foreign communication skills.

Predator Awareness is such an effort to converse to our

benefit with the animals within the ranch's boundaries. It's a conversation of postures, actions, and perceptions. This conversation is not a thinking language but an engagement of instinctual responses.

Above, we discussed our actions and the stocks responses to our engagements.

Many of the naysayer's comments are about the inability to generate the responses we are seeking.

Interactions and instinctual responses within the animal kingdom are certainly nothing new. In that same conclusion, neither are the naysayer's opinions.

Control

A natural human characteristic is *control*; we humans are comfortable with the idea of controlling and managing life's interactions and our relationships.

There are many who seek power in a guise to rule over others. Mandating other's actions as if they were ruling within a dictatorship.

That management of every detail helps establish our perception of our *control,* even when that *control may* simply be an illusion. Many times, we posture as if we are in *control,* but in actuality, we are simply a small part of a much larger picture.

The *natural world* also has an order; one could place apex predators in the *wild's hierarchy*. They are simply ruling and controlling every detail, but that thought is only a human insertion rather than a reality.

Any animal is just that, an animal. Each species of animal has established their place in the *natural world.* That social stature of ruling over others is only an illusion. The fact is all living creatures are governed by instincts. The thought of ruling over others is simply a *human characteristic* of mankind seeking power.

Seeking power

I had a cow boss in my younger years who was a very good mentor when it came to Livestock and Horsemanship. But his people skills were not so good.

Actually, when I reflected on that, there were a few that fit that scenario. But this one Boss, in particular, presented a *standout event* that proved to shape my outlook on life.

This cow boss enjoyed lamb, but that lamb had to be raised in a very specific way. These lambs, after birth, were kept in a barn for the remainder of their life. Their mothers were let out in the morning to graze, then were brought into the barn again, in the afternoon and through the night, so the lambs could nurse. The lambs remained in the barn, a shaded area, and would not be allowed in the bright sunlight.

After the mothers were turned out to graze, the lambs would be fed a ration of beans as well as water. The closest water was at a trough about a hundred yards from this barn. So, the boss instructed a laborer to feed and water those lambs each morning. The laborer would fill two five-gallon buckets and carry those buckets to the barn. The faucet filled a round cement water trough that was positioned in front of the tack room.

Many times, we would come in for lunch and change to fresh horses. The faucet where the bucket was filled pointed into the trough, so a hose was needed to fill the buckets. As the lambs were not allowed the dirty water that was in the trough, it needed to be clean, cool and fresh. Not allowing the laborer to simply dip the bucket into the trough.

Changing mounts wasn't a daily routine but simply an occasional occurrence.

One time, when we came in, the hose was still attached to the faucet; when the boss went to water his mount, the horse, being tired, drug a hind foot across the hose, simply catching the hose. That didn't get a response from the horse as it was tired and simply just wanted a drink. The Boss, on the other hand, had quite a different response. As I witnessed this, I never understood the intensity of the boss's reaction.

That small event set into motion actions that went on for

about two weeks. No matter what or how the laborer did, it was met with severe criticism. That fella couldn't even carry the buckets correctly.

This laborer was living above the tack room and bunked with another laborer; their main job was irrigating.

I received a call from the boss that I was to check some specific fields. I did so and reported back to him on my efforts.

This arrangement lasted about three years as the irrigators simply quit unexpectedly. Leaving the boss, the job of irrigating.

The boss invited my wife and I to dinner with his wife. We enjoyed the lamb; unfortunately, neither my wife nor I are lamb lovers. But we enjoyed the evening and had a discussion on the trip home about priorities as well as our interactions and their consequences.

The effort of control, or that "do as I say mentality," often hinders productive results.

Many times, I see this action when others are engaging with animals. Their solution is often pushing harder or yelling louder. The animals simply can't leave as the irrigators did; their response will be a "fight with response" rather than a "work with effort."

Such engagements, as with the boss and the irrigators, are avoidable, as are our efforts when working with animals. We should have a realization that *our actions are what engage their responses.*

Often, when we experience a problem when working with those animals, it may be a good time to reflect on our *interactions.* Rather than back stepping to the solution of *"what's required here is a bigger hammer."*

One may assume that when I speak of *our interaction* it's solely intended to our stock and our *Stockmanship skills. The predators are also affected by our interactions.*

Often, some question my explanations of our engagements with the predators. But many times, when you hear of predatory conflicts, those conflicts may often be to the same group of cattle as well as the same repeated rancher.

I realize the response of engaging the threat, I understand. But every attempted engagement is a conversation whether you engage the predators or not. The effort develops a conversation, whether it's intended or not. Those interactions are a conversation of responses that will affect the predators. It is not a face-to-face conversation as humans would have, but an understanding of your presence.

Their illusiveness means that if by chance we see them it's simply by chance, like the roll of the dice.

Their illusive nature probably means that they were well

aware of our presence.

The engagement benefits the predators understanding of our inabilities and develops a level of comfort for the predators.

This interaction is the conversation. Removing *fear and replacing it with understanding. Understanding equal's comfort; comfort encourages presence.* As I pointed out about *control,* often we humans believe that we can *control* every situation. That mindset is that "*do as I say*" effort. Forcing our will into the interaction.

My theory is that uncertainty conjures up instinctual responses. By stepping away from the superiority of control and allowing instincts to dictate responses, our interactions develop a "work with" engagement, whether it's our *Stockmanship skills* or communicating with the *predators* through an *interpreter of instinctual responses.*

Those interactions, whether with our stock or the stock's engagement with the predators, affect all responses as well as their actions.

Promoting a "work with effort" that suits our needs and benefits our bottom line is the objective.

The Community

When one thinks of their *community,* we often think of a town that we are in close proximity to. But in reality, where we live and work is our *community,* isn't it?

Well, for some of us, that *community* is the ranches parameters. Our interactions are with the *natural environment.* For many, what we're *talking about is a foreign language.* In that same light, much of their *communities'* interactions have me scratching my head. But each of our interactions within our *community* is important to each of us.

I realize that there are some who are saying to themselves, well, your community is about the people you're around. What about your occupation or even the lack of one?

Many may wonder, lack of one? Well, if you think of the homeless crisis that we're all witnessing, those communities could be thought of as communities within our community.

One's occupation and its relationships occupy much of one's time. Even if you work forty hours and have the weekends off, much of your life is part of that environment or that community. That example would equate to the homeless community as well.

I realize that there are many who look forward to the workday ending or the weekend and the non-work setting. Possibly, those people may endure doing a job that they dislike.

For many, the *ranch* doesn't fit that scenario. Don't get me wrong, there are many things that frustrate Ranchers or even anger us. Many times, those frustrations are directed by outside influences, not issues that routinely confront us, such as weather.

Many times, the issues that anger us are those issues that humans place on us that affect the marketplaces or even our interactions with the *natural resources.* But the ranch is part of us, or better said, we are a part of it. Our *community* is the ranch.

A lot of the time, many don't realize our connection to the seasons or our time-sensitive required management efforts. When it's time to do something, for the most part, there is not the option of, I'll do it later.

Many Ranchers that I've known don't abide by a clock or a weekend. The weekend may be a destination for an event of some sort, but often, those events relate to their occupation as Ranchers.

Their community is the ranch.

When you look at an urban area, there are as many different pieces to that environment as there are interactions. Whether it's desirable interactions or problematic behaviors. The individual is a small piece of a much larger puzzle. The ranch is as equally complex; our objective is to interact with the *natural environment* to promote and enhance it.

Imagine within an urban setting, one resident looking out for and protecting as well as caring for his or her neighborhood. That action would be unique, but the *ranches community* engages *in that manner daily.* We call this *Active* Stewardship.

Often, we are busy with the details of business and the efforts of production. Added pressures and new problems are just annoying circumstances that interfere with our production efforts.

We are often presented with a problem that wasn't expected. Our responses are often reactionary rather than a planned effort.

Predator Awareness is best when it's a planned effort rather than a reactionary response.

Preparing rather than reacting.

Sustainability

Sustainability is about directing our actions to the *betterment* of our future and to our community.

I realize that there is a popular word, and I often hear that word attested to as a *solution.* The word is co-exist; in my opinion, there is no value to the word. The word and its intentions were founded as an effort to unite mankind and its many different religions in 2006.

Mohammed Abdul Latin Jameel founded it in London to mitigate tensions between religions. Its simple interpretation is "Can't we all just get along."

Mankind's history shows the inability to do so.

The *natural environment* is intended to do the same. Predators require prey. *There are many who promote the idea only to the effort and not to the conclusion.*

Their efforts are profit rather than any resemblance of *stewardship.* My point is the co-existence is a word with no comparability because neither humans nor predators give a hoot to its meaning.

I wonder how many volunteers would lock themselves in a cage with a hungry African Lion and try to convince the Lion they can simply co-exist. I venture to say not many.

So, I don't like the word co-exist.

Predators do what they do; there is no way to change that; they're predators. Understanding that you also must realize that they're animals. All animals are governed by instincts as well as behavioral characteristics.

Interactions of any kind have responses, even if we choose not to recognize them. That non-recognition is a response in itself.

Ol' Smokey

My opinion is that a bear is born a bear, and that bear will die as a bear. It is just that, a bear.

Implying some human characteristics upon that bear, such as placing a Mountie's hat and a pair of blue jeans on it, may invoke some resemblance of care for the environment to humans, but it's just man's insertion of an advertising campaign intended for profit. It's definitely not a connection to the bear.

The intention of such an advertising campaign focuses on the accountability for the destruction of these massive fires on the *general public,* rather than the truth of mismanagement, set in motion by the environmental community.

Our disconnection to reality has been done with the engagement of entertainment. That entertainment is done for profit rather than any benefit to the *natural environment.* We insert human behavioral characteristics onto the animals.

In order to manage an animal, such as a predator, we must interact with the behavioral characteristics of that animal rather than applying a *human's* view of marketing. We must address it on an instinctual level in order to manage their responses.

These actions could be thought of as a conversation.

Those conversations may or may not have beneficial results,

but in my experience, there are two choices: the *work with,* or *do as I say* method. In my experience, one method works, the other doesn't.

Chapter Seven
Understanding Fear

When the instinct of *self-preservation* is engaged, it's because there is a perceived or impending threat.

(Cattle in a *herd awareness* posture)

Instincts engage a response of *fear*. Such an engagement is not a thought process, but rather an uncontrollable response. It's a response that must be discouraged and the *standing solution* must be practiced to be an effective deterrent.

Many say that *fear* has two reactions: *fight* or *flight.* Myself I believe there are three when you consider herd animals. The herd group itself diminishes the risks to the individual, which is a basic requirement for the predator.

That herd posture also engages the *self-preservation* instinct within the predators. The herd posture summons an instinctual response, challenging individualization and creating doubt. Doubt is the first step in establishing *fear*.

If you observe herd postures, whether through the Internet or in person, you will find a distinctive difference within those herd groups. Herd groups that have had predatory pressures will be much tighter and also have a heightened awareness. A nervous and unsettled herd group.

Groupings of livestock that have not had an apex predators' pressure will be a much looser and uninterested group. Unaware and carefree within their surroundings.

In a highly Coyote populated area, the stock is often dispersed evenly across the range. Not by a pre-planned effort but by the *predatory imprinting* and the desensitization that the Coyote creates. This pre-established loosening is an opportunity for an apex predator.

With the wolf's constantly expanding territories, the Coyotes have set the stock up for the Wolf. Presenting individualized opportunity just about everywhere the wolf goes.

Our stock, from that *predatory imprinting* put in place by the coyotes, have become comfortable with a *predator's presence.*

Challenging the predator's requirement of individualization of prey victims is the purpose of *Predator Awareness.*

An animal's perception of a threat can be factual or purely a mistaken circumstance. The response of *fear* is not analyzed; it is simply engaged, and the reaction follows immediately.

It is that *instinctual flinch or ducking action* we have all seen or ourselves have experienced. The response is instantaneous; it requires no thought; it is just an instinctual response.

By understanding *fear's* response, we can present situations that bring it forward intentionally, or we can avoid its engagement entirely.

By *practicing,* we can encourage a preferred response.

By recognizing that the stock is unprepared for the wolf, we can instill a preconditioned response. That response will be a heightened awareness, as well as a response to seeking the protection of the herd group.

The herd grouping of our stock today is just a group of cattle.

It has no defensive purpose.

By simulating a predator's pressure, we engage in managing our stocks *response to fear* as well as re-establishing an awareness of their surroundings. Preparing the stock in case of such an encounter and adding a solution to such pressures.

Re-engaging an instinctual response.

Critters react

Flight is escaping a *fearful* situation, whereas *fight* is engaging such a threat. The responses are each directional, running away, or challenging. Both are void of thought and are just *reactionary*.

For animals, *flight* is, in most cases, the first reaction of *fear*. The response of *fight* usually is because the response of *flight* is not available.

With our effort, we can establish, through desensitization, from the *pseudo predators,* an effective deterrent, the *standing solution, and the defensive posture of the herd group.* Interrupting the individualization that the *predators* require.

Pondering Fear

When we think of *fear*, we think of something that may have happened in our own personal past. A recollection of a scary situation, an uncomfortable memory, or even an injury. We recall the memory that still has an impact.

I realize that human societal pressures try to promote *fear* for some type of profiteering, but the level of *fear* I am speaking of is not a *fear* that someone has to convince you of. Rather, it would be a gut-wrenching response to a personal event that's burned into your memory.

Such a level of *fear* can be a self-inflicted response. Many times, in humanity, we would call such responses *phobias*. *Phobias* are often considered unfounded and a self-conjured up extreme. But no matter one's perception of another's *phobia*, the fact is that the one dealing with that *phobia*, that *fear*, is real. Realizing the serious conditions that any *phobia* presents in human behavior, we can relate to what an instinctual response to *fear* is. As with *phobias, fear isn't a reasoned understanding; it is an uncontrollable response.*

When we relate or transfer our thoughts of *fear* toward animals, we often put ourselves in the position of the animal.

We qualify the intention or the apparatus as something that would certainly scare me if I were that animal. Qualifying what should scare that animal as a first position.

Those positions are based on our perceptions; the problem is our perception is tainted by our experiences. Rather than guessing what would frighten or scare the animal, a better way to think of a deterrent is how do I engage an instinctual response?

By engaging an instinctual response within the predators, we create a reaction, and we successfully engage movement.

We do this by placing a non-understandable stimulus, or by interrupting the predator, prey relationship.

By guessing what may scare the animal, we put ourselves, our experiences, or even our presence into the equation. A better question is, how do I present an un-resolvable doubt?

If our actions cause desensitization and remove movement, we're then presenting a non-threat, which, for the predator, becomes a comfortable place to remain. Rather than presenting a deterrent, our actions actually have the opposite effect.

Contact

When we unintentionally present our presence to predators, what that creates is an understanding for the *predators* that we are a part of their environment and our presence is normal.

That normalcy equates to comfort; such comfort is to the fact of how easily we're eluded.

Not presenting our presence seems an unattainable occurrence, but it's simply a matter of a timing issue. *Predators* will be active in the early morning and in the cooler temperatures of the late afternoon, early evening, and even the night.

These times are also the times when our stock would be spread out grazing; by installing *predator awareness,* our stock will have a prepared response.

The stock's grazing times and the *predator's* presence may be a coincidence, but when it comes to *animal behavior,* I don't believe in *coincidences.*

By addressing the time we present our presence, we limit our desensitization of the predators.

If you're monitoring or simply riding to check herd health, the afternoon may be a better choice. The cattle will be shaded up and resting, all concentrated and relaxed.

This posture and relaxed position may also be a coincidence, but as I just mentioned, I don't believe that to be the case. This time offers a settled group of stock, a group that's condensed and easily located, as well as presenting the opportunity to observe the herds' health.

That timeline would minimize the chance of accidentally contacting the predators and promoting the predator's desensitization.

An unresolvable sensory perception, such as an unfamiliar scent, continues to present a question. *The questionable situation and uncertainty are as much of the deterring factor as any action.*

Once again, our actions

Action without consequences equals comfort.

We may assume that an early morning check, where no sign of a predator was seen, was a good check. Although we believe that we didn't see anything, the elusiveness of predators more than likely gave them an opportune seat in viewing our presence.

The unknown is one of life's greatest *fears*. Humanity's need for a resemblance of control is to explain away those *fears*.

Within nature, there is no such explanation; nature is only confronted and responds with instinctual responses.

The only thing that resembles explanation within nature is desensitization.

Repeating actions create desensitization. Desensitization equals comfort.

Fear Manages actions

Humanity has a long history of creating and using *fear* to manage and control the population's actions.

One doesn't have to look very far to see the manipulation of *fear,* from fines to incarceration, from social media, and to the non-compliance of whatever society deems relevant at any given time.

Societies even present their opposition to another society's actions, deeming their actions acceptable or not.

My point is that we often relate *fear* as a *fright* or *scary situation,* but in actuality, society manages our actions on a daily basis through the manipulation of *fear.* Creating compliance rather than resistance.

Fear is manageable; you don't think of fear in that context, but it can be used to manage and control a situation or an action.

Gauging Fear

Within our language, we have many words that describe similar concepts. We call those words synonyms. To list such a long list would be a pointless challenge and the list wouldn't have any correlation with anyone else's opinions.

When we are talking about temperatures, we have a gauge or a tool to measure with; that tool is a Thermometer, and its measurement is a degree. When we are trying to measure *fear,* we only have synonyms, no gauge or measuring device, only our perception and choosing a word to relay our measurement of the degree which we believe fits the situation.

This attempt will be to apply a structure to grade a level of *fear* to each response. The lowest number will be the subtlest *fear, and* the most profound will have the highest.

Each measurement will try to be tied to the *instinct of self-preservation* and the response to *fear.*

Although the lowest grade *fear* would be our first level, it would be the unknown.

1. An un-resolvable question, that simple bump in the night.
2. A historical event, a bad experience, an injury.
3. The *fear* of death. An attack or threat of a superior opponent.

Durability

Each level of *fear* intensifies and compounds its strength and its *durability. Durability* could be described as a lasting strength of a *fearful* memory. The *durability* maybe more of a concern towards domestic animals rather than to the wildlife.

When we understand *durability* as well as the reason, we require continued effectiveness. We can then realize that understanding, erases *durability, as well as removing fear.*

We install the *fear* within the wildlife and we create it for a specific purpose. That in itself is the deterrent. The installation of that *fear* suits our needs. The *durability* is the objective in order to remain an effective solution.

As to our domestic animals, we try to avoid *fear's responses.* Often, when our domestics carry it from a previous experience, it can be a challenge to overcome, as well as a challenge to work with. It's an installation that can be set for life.

When I say its installation, many are probably wondering

'installation?' If you think of *fear's* responses, you need to realize that it's a primary building block of a reactionary and instinctual response; it's a foundation that other actions are based on.

The simplest action, when *fear* is present, changes that simplest action. Something as simple as a walk, the element of *fear,* can change the cadence of that pace.

Strolling along

Think of going for a lazy stroll in the park, relaxed and at an easy pace. During that stroll, you are mugged, and you suffer injuries. Will you ever be able to return to that relaxed, easy pace? Or will that *fearful* memory change, that relaxed walk forever?

Keeping you constantly looking around for suspicious behavior.

Thinking about that stroll, what we're talking about is a reactionary response to a bad event. That experience establishes a foundation of *fear*. Your future actions will be affected by this memory.

Preparing

What if, prior to the event, you had some type of self-defense training? In my experience, the instructor would teach being *aware* of possible threats as well as your surroundings. By having a level of self-awareness and a defensive plan in place, you may have been able to avoid that mugging.

The thought of preparing prior to the need seems to be a simple concept. By instilling a response, you also establish an *awareness* prior to the initial engagement. This idea of preparing for that 'just in case' is the basis of *predator awareness.*

Stepping forward

In presenting a Solution for our stock, we can think of it as a threshold that we cross over to reach the other side in order to re-establish a response that becomes a safe haven.

On one side of that threshold are *fear's* responses, and on the other side are the *Standing Solution* and the *Defensive Posture of the Herd Group*. By moving the stock across that threshold, we provide a safe zone for the stock.

Moving the stock across that threshold is practicing a response to a predator's pressure. *Preparing* for such an encounter.

Giving the stock a solution to such pressures, instilling the defensive posture as a comfortable *safe zone.*

The key to installing this response as an effective deterrent is instilling the *Defensive Posture* prior to the predators.

Understanding a device

Explanation is the removal of *fear*. "Understanding one's *fear* is the first step in overcoming that *fear*."

The wild kingdom offers no explanation, only the animal's perception of the moment. Without an understanding, a response is unavoidably engaged. Any physically repeated action is recognized quickly, and that recognition could be equated to understanding. Even if the device creates *fear* in the initial contact, without consequence, its repetition begins desensitization, which then quickly becomes curiosity.

Curiosity is a draw, the opposite of deterring.

If we present an action or an engagement that has no repercussions, that's an invitation to a curiosity that simply must be looked into.

Repeating actions are desensitization, which removes movement as well as *fear*.

Fear's recall is powerful and long-lasting and may, with time, become greater as a recall than the actual fearful event was.

That's *fears durability*.

When we begin to think about flight, there are many reasons that could generate that response. Of course, it is the response of *fear*, but what brings forward that fear can be quite a long list. *Fear* is a powerful life lesson. We humans are bombarded daily with suggestions of *fear*, "Do you have enough insurance"? "Is your home safe"? "Are you eating right, do you (fill in the blank)?

Because of this daily bombardment of news and advertising, we as humans may discount *fear's* responses and its actual reactions.

For no other reason, we believe those responses can be ignored just as ignoring those irritating commercials.

That is, for us, desensitization, but our understanding of commercialization is why we can understand that this isn't something to generate an actual response of *fear* for us. That understanding removes *fear*. Creating for us a blurred vision of suggested *fear* versus a truly *fearful* situation.

We have become desensitized to commercialized suggestions of *fear*, but true *fear* certainly engages actual responses, such as with phobias.

Because of our societal desensitization, our judgment of *fear* and what should present it may be compromised.

Fear is not a pillar of some indestructible substance. *Fear* can be overcome. A realization is that if something remains mysterious or not understandable, *fear* remains and may also increase its intensity.

Confusion, when engaging *fear* for management practices,

is a strong ally. When we understand that *fear* is manageable, we can incorporate those practices into actions that affect animals. Whether it's our stock or the predators, effectively managing each of their responses is our job to understand their behavioral tendencies and species characteristics.

Fear engages movement.

Desensitization removes movement as well as *fear*.

Often, we humans equate *fear* with being scared or even frightened. Instinctual reactions to *fear* aren't frightening, there simply not understandable.

If the task is removing *fear*, understand you must engage a repeated action to ensure those actions are understandable, desensitization.

If you encourage reactions of *fear* to your benefit, you cannot allow understanding.

Domestic versus Wild

The Natural World may seem like a fun and inviting place, much like a cartoon where all the animals are always frolicking and happy. That Hollywood version is far from the truth. It may have sold many movies, which was the intent, but it certainly isn't reality.

The Wild is certainly that, Wild. The *instinct of self-preservation* is the strongest instinct for a good reason, survival.

This is why wild animals are so elusive. They're elusiveness is from living with a daily *fear* and simply surviving.

The wild isn't focused on *fear,* but each animal is aware of their vulnerability. If the predators establish such *fear* within our stock, not only is that a bad experience for them, but it sets to memory a foundation on which other responses are based.

When *fear* is injected into our stock's reactionary foundation, that same survival mode that causes wild animals to be so flighty and elusive, makes our stock nervous and difficult to handle. As well as the cattle's demeanor it presents a foundation of stress that effects their performance.

An animal's response is reactionary; their perception is just that, their perception.

When we think about an animal's response to a perception, that response will be linked to past experiences and *fears durability.* Previous experiences can establish severe responses.

Wild Mustangs

(The following discuss's *fear, fear* crosses species lines. I use horses as an example because they are better and more easily understood than predators.) When we're talking about horses there will be quite a difference in a horse that is raised in captivity versus one that was raised in the wild.

An example: maybe two young horses just beginning their working relationship with you. One was raised in the wild, the other in captivity.

You're riding down the trail and a butterfly fly's alongside you and towards your horse's head, possibly in the direction of the horse's eye.

A response from the youngster who was raised in captivity maybe a snort and step away from the presence in recognition of the butterfly.

Whereas the youngster that was raised in the wild, will have a much more severe response. More than likely its reaction will be 100% *fear*. With every ounce of power that youngster can muster, it will jump to the side, followed by an all-out run, which is flight.

That young Mustang's response has left many of us lying in the dust and dirt, scratching our heads as we're gathering ourselves up and telling ourselves, well, that was dumb; I saw the butterfly coming and didn't connect the dots. This comparison of a domestic animal's response to *fear* versus one that was captured in the wild, is intended to illustrate the wild influences on an animal's responses.

That response and that wild individual's reaction will not be a thought process, but an all-out instinctual reaction.

This response and understanding, is a useful tool, whether deterring presence or mitigating the risk that predators present.

All youngsters of any species can have a similar response when they're relaxed and just walking along. When I was a young lad they told me I was day dreaming.

For the young horse or whatever the animal, it can be just an idle mind.

The Boogie Man

The idle mind invites what I call a *boogie man event.* That *boogie man* was just waiting to spring out and engage *fears* response. Just as with the butterfly these responses are quick and strong.

Boredom and redundancy can quickly turn into that *boogie man event.* Often as riders we become relaxed and complacent because our youngster is doing so well. Setting the table for our guest the *Boogie Man.*

As riders, we can keep the youngster's mind occupied to prevent the colt's mind from slipping into a *fear* and *boogie man event.*

An example of keeping a colt's mind occupied maybe pace, working on slowing down and speeding up. It's an opportunity to work on subtle suggestions, shifting your weight slightly forward to extend the front foots reach and its placement. Or sitting down to begin slowing and finally inhaling and moving your heals slightly forward, relaxing into a stop.

When you have a youngster that's a traveler and he's always crowding pace, a suggestion is allow such a pace and create an action that works with that desire. This action of crowding is often present when your colt is tracking or trailing stock.

A good action would be a zig zag pattern rather than a straight line. Working on turning and the hips disengagement. Developing feel and preferred responses rather than trying to hold the colt back. *Working with,* rather than *do as I say.*

When we interject this method of the *boogie man* as a deterrent, we present an unknown into a very familiar place. That unknown is much like a child's *fear* of what's in the closet.

We need to understand that *fear* is powerful, its responses are not analytical. They are completely reactionary. Are they always predictable? Of course not. The animal's perception is their own.

I have found tracks where I had previously scented. I also have heard comments from people discounting scenting because of those tracks. I myself was left wondering.

A predator tracks in the area may have had other reasons for occurring. First is the possibility of the chosen *scent deterrent* being simply a poor choice, or simply misapplied and simply became understandable. I will keep that scent and put it on my *do not use list.*

There may be other reasons, such as the predators themselves.

Their mind set at the time of the *scent's* engagement may have been occupied, intent on another objective, such as stalking or chasing an intended prey. The occupied mind may not allow an interruption in action, just as our action of occupying the colts mind, which is intended to prevent such a *boogie man event.*

The predators may have a similar reaction, simply being engaged with the task at hand.

In my experiences, I have not found an engagement sight. I realize that every engagement isn't always successful.

Possibly the deterrent took time to interact with the occupied mind, creating a delayed response.

For the young horse and the *boogie man event*, an actual stimulus isn't required. The idle mind just creates it, and the engagement of *fear* just happens. By occupying the colts mind with a chore we prevent the *boogie man event, the occupied* predator may block the instant recognition of the deterrent as well.

In the case of presenting a deterrent within the wild kingdom we are actually placing a stimulus to create the *boogie man* and present the unknown. The response may be different between individuals, just as with humans, what may scare one severely, maybe a modest fright for another.

The unknown is key to the *boogie man,* whether a young colt or when we place such a deterrent to deter predators. Neither the colt's timeline nor the predators are predictable, but *fear is an instinctual response rather than a reasoned action.*

A Difference

The individual horse will even be different within domestic settings. Let's say you have a small farm that has a couple of mare's, which you raise and then sell their foals.

Those foals are in constant contact with human interaction and have been rubbed and worked with since the first day. This is called imprinting. This young horse, if taken in the proper direction, will be a willing and easily developed individual as there is no *fear* within Its foundation.

This imprinted individual could also be equated to a young predator. If a youngster realizes that humans are simply part of its community, they will become comfortable within close proximity to humans.

A larger and more substantial ranch will have larger ranges and larger mare herds. Their experiences will still have contact, but just less of it. Their experiences will still not be linked to *fear.*

The abundant contact of the imprinting will be an advantage for the one colt, but for the other it will simply require more time.

The Mustang colt will have no human contact, no imprinting, only *fear.* In the wild, the horses are vulnerable to predators and humans, although, with all of the best intentions, the wild horse only perceives that the human chasing him isn't any different than the lion who just chased him a couple of days ago.

To the Mustang, the Lion as well as the Man are both predators.

Being raised in the wild, survival is paramount. *This fact means that fear and its responses are part of the daily routine.*

Although we are speaking presently about horses, I just want to point out that this fact is prevalent in all wild animals whether it's horses, game or the predators. Awareness and familiarity establish comfort which provides understanding. No matter the age of the individual horse within a herd, that herd response is training the foundation of *fear* within each individual horse.

Understanding that all animals can be affected by *fear,* we can then attempt to manage it.

Realizing *fears responses* can put reactions into context. If we desensitize, even if its unintentionally, that understanding removes *fear*. Providing a comfortable place to remain, promoting a predator's presence. Comfort quickly becomes habitulization, which begins with understanding.

The Wall of Fear

If we think of each *fearful* event as a wall that continues to rise with each *fearful* situation. Think of that wall as elevating the responses and intensifying each reaction. Giving wild horses in comparison to their domestic counterparts a much greater reaction to perceived threats.

When we think of any animal from the wild, our first understanding should be that they will be extremely reactionary in any response.

This reactionary response is a foundation of the self-preservation instinct; it's why engaging an instinctual response is an effective deterrent. By understanding this established reactionary response, we can present it and manage it to our benefit.

When working with the wild horse, their individualization only compounds their *fear*, making their foundation built on reactions rather than responses of applied pressure.

The individualization I am speaking of, is when we are working with this wild youngster. We are simply a predator in the young horse's mind. Our attempts of working with this individualized, *fearful* youngster, isn't any different to it than a predator chasing it around the pen. The challenge then becomes engaging trust. In order to reach that trust, we must reach the other side of that *wall of fear.* That's the threshold we are challenged with. The established *fear* is a challenge to overcome.

The goal with beginning this wild horse is different than the domestic youngster. That domestic youngster's previous experiences don't exist with this wild individual.

Actually, the wild individual is in a much lesser advantage because of its established level of *fear* and experiences which have established that walls height.

The simplest basic effort in training for the wild horse will be first in developing a trusted relationship with the individual. That may sound like a simple task, but the degree of *fear* will dictate whether the job can even be accomplished. In understanding that trust isn't simply given, but rather it must be earned. We're able to comprehend that wild animals are just that, wild. We simply can't install trust. That understanding qualifies the value of managing fear and its response's.

Providing understanding, removes *fear. Our presence quickly develops the predator's enhanced elusiveness and promotes understanding through our presence. Developing and promoting the predator's elusive nature.*

Rather than deterring, with routinely practiced presence, we are unintentionally building a relationship of trust for the predators. Promoting presence, or as some prefer saying, coexistence.

Building that wall

We have just talked about *fear* and each event being a building block in that wall of *fear.* With each engagement the *wall* strengthens and continues to rise.

When we think of that *wall of fear,* we could think of it as a physical wall.

Simply getting by that *wall* may not be possible.

That *wall's durability* may be a long-lasting *wall* for an individual, or it may be simply a minor detour in another individual's routine.

Many times, in our own journeys, a detour may not change our intended route. But, multiple detours and their engagements, eventually has us looking for an alternative corse in reaching our destination.

That repeated *blocking*, just as with our detour, requires changes and requires establishing an alternate passage.

Establishing the importance of keeping the deterrent presenting a new and routinely engaged *wall.*

Comparisons between species makes the response of *fear* more understandable across species lines. *Fear*s responses are durable and effective as well as powerful.

As in *scenting,* by changing to a new scent, we continually present a new and unfamiliar unknown. Each encounter will be adding another building block to that *wall of fear,* elevating that walls height with each engagement.

For the predators each new interaction that elevates that *wall of fears height,* makes each contact more reactionary.

Whether it's discouraging presence or challenging the herd group, building that *wall of fear* is about developing *a response rather than an action.*

By recognizing that fear is manageable we can realize an effective solution.

Trust

Gaining *trust* doesn't start with a clean slate. That wild horse that we are working with may have a long and serious list of pre-structured *fears.* Gaining their respect first and then moving to *trust* where simple training may begin, may not be possible.

I don't suggest that every horse is a hopeless effort, but rather each wild horse will require more effort because of its pre-existing experiences.

As for *stockman,* acknowledging *fear* has been a discouragement. It's a common statement that "there is no place for *fear* when working with animals." *Fear* is a natural response; it's base is deeply rooted within all living creatures.

Understanding it, provides alternatives to its negative effects.

I would rather encourage developing *trust,* as well as understanding *fear* and its responses as a preferred *Stockmanship skill.*

Subverting fear

Stock that has had previous predatory pressures, will have a much quicker and reactionary responses to any perceived pressures. The pre-established *fear* promotes an unsettled and nervous demeanor. This unsettledness effects the stocks sense of well-being. That unsettledness effects fertility, performance as well as their handling.

Gaining *trust* after *fear* has been established in the stock is a challenge in of itself. The stocks response to any contact quickly escalates to *fear.*

In my experiences, the challenge for stock that have been predatorized is one has to *desensitize* the stock. By *desensitization* you remove movement as well as *fear.*

The settling of the stock will have to begin where containment can be controlled. Slow and steady movement in a lateral pattern outside of the flight zone will slowly settle the stock. But, in my experiences this will take more than a few sessions.

What's important to realize in rebuilding *trust* is that the mechanical portions are important, but not the most important. The most important point is the *release. Pressure and release,* just as we use in training, such pressure has already been established by the predators. Our goal is *obtaining the release.* Each time we quit and walk away, it provides a release.

We don't have to escalate the pressure; we only have to work on providing the stock with the release. Our presence, or any perceived occurrence equals a pressure.

By affording our stock with a *calm and uniformed stance,* the stock become *settled and attentive* to our presence. By allowing enough *dwell time* and then *releasing or quitting at* this time, we provide an instruction of where to find such a *comfortable position.*

This simple *re-settling* of our stock is a *challenge; that wall of fear* is the opponent. If such *fear* is too high, correcting or reinstalling *calmness may not be possible, just as with the wild horses.*

Balance

We talked before about there being a *balance in nature* and *nature always seeking its equilibrium, s*uch as water seeking an equal level or that simple *scale balancing.* With that thought, any action and its responses also have a *balance.* Thinking about that simple scale balancing, if we were to place a finger on one side, our *pressure* would cause the other side to rise.

If we then removed our finger, that *release* would allow the scale to return to its *balance. (Pressure, with a Release, equals balance.)*

Behavioral responses are much like that simple scale. Any pressure has a response.

We humans are analytical and require a thought process, or do we?

If you think about going to the Doctors for a check-up and that Doctor sits you on a table to check your reflexes, the Doc taps you with the little hammer on your knee and your leg jerks.

Was that jerk of your leg, a thought process, or simply a physical response to an applied pressure?

Because of a human's analytical thought process, we would simply say that's just a reflex. As humans we have a need to categorize specific actions with specific words.

But one could also equate that it was a *response to an applied pressure.* After that response or reflex, your leg simply returned to its relaxed status, *no worse for wear.*

Would we call that response a *release, or a Balance?*

When we think of a *release,* we think of an equated time frame, don't we? We also think of a time frame of *pressure,* especially in working with animals. If we go back to that reflex of the leg jerk, the *pressure* that caused that jerk was the contact of that hammer; the actual strike would be measured in milliseconds, and the strike itself equals the pressure. The *release* would be the instant of the completed strike.

The *balance* would be the return to the relaxed status.

Pressure with a release equals balance.

Often with *applied pressure,* the *response* is a quick and immediate reaction.

I'm sure many times we create that knee jerk when working with animals, just as the Doc's strike was only a brief contact, our action may have been an unintentional engagement of our stock which created a response.

This immediate response often generates a reply of "*I didn't do anything*".

In actuality that may be true, it may simply be the animal's perception of that moment, rather than any physical stimulus.

Simply a *boogie man event.*

The *response* doesn't always equate to *actual physical pressure but* can be a response simply from the animal's perception, much like a flinch or that ducking action.

The *challenge of balance* is an awareness of *applied pressure* as well as recognizing an animal's postures and realizing a rising level of *tensity.* Understanding these elevated *tensions* are where *good Stockmanship and Horsemanship as well as dogmanship* begins.

There is a fine line in defining the line between *too much pressure* versus *not enough.* The animal's response is a *reaction* rather than an *action.*

Any perception can be magnified by any level of questionable circumstances. Even if we're disconnected from the event that instilled such *fear,* we can apply that theory of that scale seeking its balance, as a solution to that *fear mindset.*

The question then becomes how to create the release.

Finding Balance requires reestablishing the release

Understanding the mechanics that are a combination of *pressure with the release, which then engages balance.* Is in of itself, the explanation.

Going back to the *reflex,* the Doc's applied strike was the pressure. There is no way to take that applied pressure back, the applied pressure has begun an action. That action or reflex just has to play out. Correcting an instilled *fear* can't be reversed. It must be allowed to work through all of its required steps, *pressure, release, and balance.*

The *pressure* after a *fearful* event may simply be your presence. Presenting your presence as a repetition of non-pressuring movement, will begin *desensitization. Desensitization removes movement as well as fear.* Then by stopping and allowing the dwell, simply standing you begin that *release of pressure.*

Through these steps and their repetition, you will slowly re-establish *balance.* We discussed that simple scale and its responses, pressure, release, balance. Sometimes *fear cannot* be overcome. The *fear* is just too heavy for the scale to handle, simply breaking the scale.

Any pre-established *behavior* is a challenge to overcome. We chatted about that wall of *fear* and its challenges. We also talked about that threshold and moving our stock over that threshold. On one side of that wall is *fear,* the other side is the *defensive posture of the herd group* and the *Standing Solution, balance.*

As we stated with the *wild mustangs* that wall of *fear* may be simply too high to overcome, but for our stock the *herd group* itself has a calming quality.

Aiding us in the recovery from the installed *fear.*

Responses are Curious

We have chatted a lot about effecting an animal's *response*. But a more curious *response* is about the human feedback.

I have been helping some Ranchers in deterring wolves, successfully I might add. Their neighbors and others, have suffered losses, I offered my assistance and they simply were not interested.

I had even found depredations on their range and got the responses of "our ranges are no concern of yours".

On some of my leases, where I have my cows, I have had neighbors that had predatory losses and I informed them. When they finally made it out to check, they couldn't find any evidence. The curious thing was, when they checked it was two to three weeks after I informed them of the loss.

When their stock was simply hard to handle, the stocks demeanor was of no concern, as everything just seemed normal. The only thing that was needed to be done, is just *cowboy up and getter done.*

One of the comments that I often hear, is that *the predators* will soon learn and adjust to our deterring efforts. That statement is assuming all of humanity's responses will be to adopt these solutions.

I'm actually confident that won't be the case, it's simply not a behavioral characteristic of humanity.

In my experience, the Ranching community has an unmatched level of independence. Those independent operators have solutions that are predetermined mindsets.

Many of those conclusions are based on previous experiences. Often those experiences haven't been beneficial to them. Most of the times these Solutions are forced upon them and those solutions often are taking some piece of their operation away from them as the remedy.

Such as water rights, or the ESA, or Air quality, or NEPA and the list goes on.

I am confident that *managing fear* will not be duplicated on a wide spectrum. As mankind's history attests to this, the few using *fear* to manage the many. That history record shows a defiant stance until such defiance is ended by regulation. The independent rancher has been routinely beaten up with solutions. His or Hers adoption of *outside opinions* is just a signal to beware.

The loss of their property rights is often the solution.

New Ideas

I remember, as a young man going to the *California Livestock Symposium,* which was soon after discontinued.

There was a fella there, that I had bumped into a few times.

His method of starting a young horse influenced me early on to that *work with mindset,* rather than that *do as I say* mentality.

I believe that I have had many mentors, some popular, some not so. *Predator Awareness* has been a culmination of those experiences and successful teachings of those many mentors.

I remember a common response to *new ideas,* has generally been some type of disqualification from the many, to such new techniques. Yet, over time many of those methods have become readily accepted and continually taught practices.

Youngsters dream

This industry has a multitude of players, all in this business, for different reasons.

I have worked on many ranches, from being just a cowboy all the way through management and, finally, ownership.

There are many different reasons for being in the business. The ranch could have been passed down through the family, or it could be a tax shelter for a successful entrepreneur. Or it can be that desire to have a lifestyle of peace and serenity. Whatever the reason, there's a very small portion of the industry that sought the challenges and the meager lifestyle that a ranch provides. Counting on that ranch to provide your income.

As a young man, I heard a speech at a yearly state FFA

convention by Roy McFee from Production Credit. He stated at that time unless you're born into this industry, there is no way that you can acquire a Ranch on your own. I was taught very early on, "Don't put me in your we." I still hear those comments today about how young people just can't get into this industry.

Challenges are tough, but a challenge is just that, a challenge. If you're interested in a less challenging avocation, build your wealth somewhere else. Then, return to the lifestyle as a reward for your efforts. But, if the industry is where you belong, that in of itself is the reward.

Chapter Eight

Controlling Movement

Just as with a vehicle, it doesn't simply move randomly; its movement is guided. *Movement is Intentional.* As to the vehicle's movement, that movement takes a human engagement.

Understanding Pressure Point Pressure

Predator Awareness also requires a human engagement, which will develop movement or disengage it.

Within our vehicle, the gear selector engages direction, forward or backwards. The steering wheel determines left or right. The brake pedal stops or slows movement. We are all aware of those engagements and are comfortable with them.

Beginner drivers are less comfortable than experienced drivers; experience is the result of repetition; any action repeated enough becomes comfortable, and that comfort could be relayed to the predators as well. When they become comfortable, it's called habitulization.

When we act as a gear selector for choosing responses with animals, just as with the inexperienced driver there is a lot of questioning and lack of assuredness. But, in the same scenario, the experienced driver was once a novice. The key to keeping any deterrent effective is not allowing the predators to become comfortable with their engagement.

For our cattle, Practice installs a reactionary response, very similar to muscle memory in sports. When we break down such a comparison, *it's simply training a response that will naturally be recalled when a situation is presented.*

In any sport, practice is just part of participating in that sport. Muscle memory is thinking of a proper action and repeatedly practicing that action. Such action then becomes comfortable, and it's not necessary to think about the mechanics; the proper action just happens without thought. The correlation is the established response.

A Coin's Value

A coin has two sides, and so does *predator Awareness.* We are managing an encounter, and each side is different, just as with the coin. The difference between the two sides doesn't affect the value of the coin. Its value is the combination of the two sides.

To manage a predator's actions, we must keep *fear* in place; we can NOT *desensitize.* As to our stock, we must re-engage a faded instinctual response, that requires practice, which promotes desensitization and promotes the *Standing Solution.*

If one were to think of another example, one could think of a magnet. Each side attracts the other, negative attracts positive, and prey attracts predator. If we were to match the positive to the positive, the two would be incompatible. Engaging a disconnect. By engaging both sides, our stock as well as the predators we establish a positive solution. Engaging instincts on both sides of the interaction. Just like the magnet, positive-to-positive actions repel rather than attract.

Predator Awareness addresses both sides of the Predator vs. Livestock confrontation. Addressing the interaction from the animal's perspective rather than trying to engage each predator with some sort of action or apparatus. Just as with the coin, the two sides are each part of the coin's value.

Driving Cattle or Driving a Car

We just chatted about a *gear selector* in a vehicle. I want to point out some similarities between *Pressure Point Pressure* and the *gear selector* in a vehicle. Through my web sight, I have had questions about *pressure point pressure.* This explanation is different, but it's position of pressure is the same. The effort is to realize that our positions and our location in relation to the animals affect their actions. As well as our presentation of direction, along with our speed, those positions dictate the animal's response.

We talked briefly about the *gear selector* engaging movement from our positioning and *pressure point pressure.* Another factor is speed control; in the car, that's the accelerator or the brake pedal.

Those two controls equal our position within the *flight zone* of the stock. The closer you are to the stock within their *flight zone,* affects their response time and the speed of their actions.

Closing that gap between you and the stock could be thought of as mashing down on the accelerator. That pressure, wether mashing that pedal to the floor or simply slowly applying a steadily increased pressure, will dictate the speed of the cattle's response.

Too much mashing of the accelerator will draw forward a response of *fear.* It's not much different than revving the engine for too long; the engine would simply overheat. The cattle would have the same reaction, overheating and engaging in *fear's* responses.

Slowing down would be applying the brake, which for our cattle would be increasing the distance of applied pressure within that *flight zone.*

I realize that we have two feet and two pedals, and we can apply pressure to both pedals at once, riding the brake. Our effort of managing movement is close enough to generate the response we want.

The speed we want is an effort of being not too close and generating too much speed but not too far away as to promote too slow of pace or even a stoppage of the desired movement. The middle ground of that applied pressure could be thought of as a constant adjustment of pressure, much like riding the brake.

The shift column has six positions: Park, Reverse, Neutral, Drive and two lower speeds of Drive. These positions engage movement; by human interaction, the desired movement is created.

We could equate these column selections to pressure point pressure in moving the stock.

Park is the eyes forward. *Pressure* to this *point* results in stopping forward movement. If the cattle are contained, such as in a crowding alley, *pressure* to this area will generate a backing-up response.

Park is also the *pressure point* of the *draw of the eye. The* draw is a question of what? But it's the stock's perception, not ours.

Reverse Pressure Point is where movement can be created or stopped depending on the direction of the applied *pressure.* If your traveling head to tail, it will generate movement in the opposite direction in which you apply the *pressure, The Reverse Parallel.* If you're passing within the *flight zone* going from tail to head, this will stop or slow forward movement.

Neutral, this area is multi-directional. Any applied *pressure* will be responded to, with a relief to the applied *pressure.* In other words, your direction of applied *pressure* will generate the response of direction. Going straight at the left rib will make the cow step to the right, *pressure* to the right and the movement will be to the left. It is one of the *Pressure Points* that have steering capabilities.

Drive, this area is home to the flank; the flank is one of our stock's most vulnerable areas to predators. That vulnerability quickly engages an instinctual response if the pre-conditioning of *Predator Awareness* isn't applied first. This *pressure point* will create forward movement.

This drive area of the flank could be thought of as a designated accelerator pedal for the predators. Pressure by the predator engages flight. This is a sought-after area for predators. As I stated before, when it comes to animal behavior, I don't believe in coincidences.

The Flank and proceeding backwards also engages forward movement until you reach the tail.

The tail itself is a *Pressure Point* as well. The area of the tail is also a vulnerable area to predators. *Pressure* to the tail will also cause the *draw of the eye; often this draw will engage an about-face.* Facing the applied *pressure.*

This *tail pressure point* is why when a stoppage occurs from a perceived pressure, and we choose more pressure as a response, the cattle will break and go in the direction of the applied pressure. Simply breaking away.

If a group of cows were *pressured* by a predator and had received no training. The cow in the last position, *the tail of the herd group, would be in that position* of the tail pressure, engaging that *eye forward connection, stopping that individual cow.* The pressure makes her stop and face that pressure. Stopping flight and engaging the fight.

Whereas the surrounding stock would be engaged in the *chase sequence. Leaving* that cow who had direct tail pressure, who spun to face the predators, individualized to face that threat on her own. Making a score in the winning column of the predators.

Individualization is a loss in our column of deterring.

Chapter Nine
Our Perception of Image

When we think about a wild horse's response to *fear,* we can all imagine the wild herd running across the range, evoking a romantic Western scene. That's another Hollywood version when in actuality, the horses are running out of *fear's* response.

A non-reactionary movement of horses would be a walk or trot. Of course, there are times such as frolicking or just an individual feeling good. But for the herd group, running is flight, an escape.

The *flight* of a horse has a measurement of time and distance once *fear* has been engaged. This measurement is a consistent response, and each species has its own timeline and parameters. How far will this flight be?

We mentioned that wild horse herd running, why does the *fear* engage the entire herd? Or possibly a better question is, how does the herd relay that *fear* in unison to all?

We also mentioned that human flinch as an uncontrollable response. Within the herd, *fear* must be ready to be released more readily than animals who aren't challenged with survival routinely.

This unison action is a herd response; that instinctual flinch maybe the same thing. Within domestic animals such as our cattle, that unison response has been weakened through years of separation for management practices as well as the lack and quantity of predatory species.

In order to reinstall that unifying flinch of the *defensive posture of the herd group* within our stock, that requires our efforts.

By re-engaging a *Defensive Posture,* we reinstate a faded instinctual response for herd animals.

When we discuss *Predator Awareness* as a deterrent for our stock, we get entangled in many opinions. Low-stress Stockmanship vs. High stress, dogs vs. no dogs, training stock or not being able to train the stock. Or the cattle are already herd animals, so there is no need for any more effort.

Proficiency in anything requires practice. Just because cattle are standing together, it doesn't automatically equate to a *defensive posture.*

Cattle that are grouped together and staring at a curiosity could certainly be defined as a herd group. But, if something caused a surprise, the reaction would be a response of fear and flight. That instantaneous reaction would not be an unifying herd response but flight in all directions.

If you think about the *defensive posture,* you could compare it to marksmanship. Just because the weapon is discharged, it doesn't necessarily mean it hit the target let alone the bullseye.

If one's goal is the target, that may happen by chance, but practice certainly increases your chance of success.

Predator Awareness is re-establishing the *Defensive Posture of the Herd Group.* It's that developing the flinch that creates a *herd response,* which requires practice.

When we think about dogs vs. no dogs, with the dogs, what we are trying to avoid is the stock being killed by a predator.

Through the years and working on many ranches, I have seen good dogs and some very bad ones. The worst-behaved dog is still nothing in comparison to a Wolf.

That bad dog, at its worst, may create some damage or even create the need for additional labor for gathering or maybe repairing fences.

Whereas the wolf will generate more substantial economic losses.

Establishing a Basic Awareness

By exposing the cattle to stock-dogs, you begin controlling the cattle's movement in ways that will assist you in management practices. The stock-dog will encourage the cattle to move as a herd group.

That use of stock-dogs, which will assist us in management practices, is in of itself a very good benefit to draw on. But yet another beneficial effect is simply presenting the K-9 species to the stock.

This response to K-9's allows the stock to begin developing a response and an awareness of outside pressure applied by those dogs. This begins a *basic awareness* for the stock to their surroundings as well as the development of a response to canines.

That development of an *awareness of their surroundings* is a *basic step* in deterring conflicts from any predator.

We had just talked about a walk in the park and being mugged. We also talked about taking a self-defense class prior to that mugging. Every instructor I know promotes being aware of your surroundings. Stock-dogs do just that for the stock; they engage an awareness for the stock to their surroundings.

Once the stock are familiar and accustomed to the stock-dogs, you then have set up the situation for the introduction of the *pseudo-predator.*

The *pseudo-predators* have a completely different agenda as well as actions than the stock-dog. But the stock-dogs have instilled a basic recognition and awareness of canine pressure.

The difference is stock-dogs manage movement, *pseudo*

predators remove movement and develop the *standing solution.*

The Wolf, or actually any predator, requires an individual. In order to create the *defensive* posture *of the herd group,* we need to show the stock that being individualized is an uncomfortable situation.

Pseudo Predators

Presenting the *herd group* as the place to seek comfort and safety is paramount. In order to accomplish this, we need a dog whose tasks address just that.

This dog moves quickly, and its basic task is removing movement. The stock-dogs, which work with a calm, stealthy type tactic, their job is managing movement.

A stock-dogs purpose is an extension of the stockman, helping in the management of the movement of the animals.

This dog that we call a *pseudo predator* is fast and encircles the stock, constantly stopping them.

(A young pup intensely waiting for the command of a send)

I have been asked by some about what breed is this *Pseudo Predator* and if it must resemble a *Wolf?* The *Pseudo Predator* does not have to resemble a wolf. It's the *Pseudo Predators* actions that are important. We are establishing a response to outside pressures.

It *must be a strong-heading dog with a lot of speed.* I have seen many different breeds that fit those proper actions. That baying of the stock isn't a preferred action for a quality *stock dog.* But it's the most important skill that the *Pseudo Predator* must possess.

I like to see some shedding ability as well.

Desensitization occurs from repetition, such as the *sacking out* practice on a young colt. Speed and the encircling of our stock by the *pseudo predators desensitizes* the stock. *Desensitization* removes movement.

This *desensitization* and removing movement are a basic principle of the *Standing Solution.*

By stopping the pressure after the stock has been grouped and shows no signs of movement and are intent on watching the *pseudo predator* encircling them. Stop the *pseudo-predator.* The key to understanding if the stock is ready for the release is if they're grouped into a divergent stance.

This stopping of the stock is the primary foundation of

developing a response to a predator's presence. By releasing the pressure from the *pseudo predator* at this time and allowing enough dwell time, the stock establish a response to such pressures. The relief to such pressures is the *herd group.*

With this response, the herd group becomes a sought-after safe zone.

This session is the primary engagement. Additional training will be required.

Many oppose it and believe it's cruelty

Everyone has their own perception of how they want things to go. That running herd of Mustangs conjures up a beautiful, romantic image of the West. In reality, the herd is simply horses running in fear.

For some, the dogs promoting the *defensive posture of the herd group* is abuse. When in reality, they're developing a response to predators that could save the stock's lives.

Whether it's the romantic image of the horses running, which is most often fear. Or whether it's promoting the *Standing Solution* to protect the stock from predators. Often, one's perception is a wrongful interpretation of a situation. For myself finding no depredations and rather finding content stock is my preferred image of the West.

How far?

When we think of movement as an instinctual response, such as the herd of horses, we can think of other events that would have had similar results and were based on a response to *fear*.

The famous wolf OR7, who made his journey from Northeastern Oregon to California, who's journey was captured by GPS. That journey was possibly a response to *fear* when he was trapped and collard.

If we realize that flight is measurable, then distance becomes the question. If we are aware of symbiotic relationships within the wild, and I'm not suggesting that Hollywood version where all of the little animals go skipping and singing to the commercial. But rather a true relationship, such as between predators and specific species of birds.

Those symbiotic relationships should also move when we use intentional *fear* as a deterrent. Taking with the predators, all of the avian counterparts who count on the apex predators for their sustenance. It has been my experience in deterring a predator's presence, that you also deter those avian species as well. These symbiotic avian species give an alert or a status update when monitoring for predatory presence.

I have left baits that would have been ravaged by birds in the past, and they have remained untouched for days. I actually picked them up and disposed of them simply because my point was made.

The deterring effort makes me believe that the predators exceed the scouting parameters of the symbiotic avian species. The effective distance of a deterrent may exceed the intention.

The key is not allowing the deterrent the chance of becoming a desensitizing device.

When we go back to our fear gauge, we must realize that we can't reach anything more than a level one as a deterrent.

What happens when we elevate a deterrent's *fear* level is that we enter into an area that is *fearful* but understandable.

An example would be us chasing the predator. The predators know why they're running, you're chasing them.

When you think about humans chasing a wolf, for us, that satisfies us personally.

For the wolf, the understanding is you're chasing it. The *fear* has been removed, just as the saying "understanding your *fear* is the first step in overcoming that *fear*."

The wolf realizes that you're chasing it; movement away from your pressure is an easy and simple response. That understanding is not *fear*.

I have heard the comment "that you need to make the Wolf think it's going to die." In my opinion, what the Wolf quickly becomes aware of is how easily your pursuit is escaped from. There is no realization to the Wolf of the seriousness of your intentions, only your inability to actually engage him.

Often, we see this desensitization occur in lethal control; the target soon becomes aware of the range of a rifle shot. Many times, when it's out of range, it will stop to see what your next effort is.

That awareness eliminates an instinctual response and changes it into a cat-and-mouse game. The *predator* may simply relay your efforts as a playful engagement.

That interpretation lends itself to the canine species.

Such interactions only establish understanding. Understanding creates curiosity; curiosity is a draw.

The effectiveness of a deterrent is the unknown. The continued effectiveness is a constant change and remains an unresolvable mystery. Engaging instincts which result in an uncontrollable reaction.

Addressing the predators with an action establishes an understanding, simply becoming a playmate for the predators.

Wildlife and Suburbia

Here within the *inter-mountain* high desert, we have two deer species: mule deer and the black tail. There is another, which is a hi-bred called the Klamath. The deer here, as anywhere else, are a desirable game species, and many humans follow their behaviors.

Often the deer will interact within suburban areas. The time is most often mid to late summer. This interaction is possibly to capitalize on the green lawns and tasty flowers within the yards but is also a possibility of the seasonal reduction of water availability.

In Northern California and Southern Oregon, I have had several wolf biologists who came to California to view our newly established residents. They told me in August that the wolf would require much more water. I explained that our water is plentiful until about July through October when it's dry. Their response was they thought the wolves would depart such a dry climate. I often heard how adaptable wolves are. This photo suggests that the wolves will endure dry conditions just fine.

(Scat from a dehydrated wolf)

The deer's increasing presence through the dry months, suggests that their encroachment into Aunt Mary's flower beds, may be because of those dry conditions. All the wildlife's parameters are condensing to the shrinking water supply, and the prey species are just feeling as entrées on a buffet line.

The deer's response to the increasing predator concentration may be getting closer to man as the answer to avoiding the predators rather than a pre-planned assault on Aunt Mary's flowers.

This seasonal migration is but another suggestion that *fear* is manageable. The encroachment into the urban area is to avoid a predator, but only to be amongst another.

Managing fear to our benefit

If we take a moment to examine the *Defensive Posture of the Herd Group,* what that presents to the predators is a question. Can I safely individualize and secure an individual without jeopardizing my own safety? Once an animal questions its ability or safety, *fear* has already engaged a reactionary response.

This question is not an analytical thought process for the predators.

The explanation of such a question is needed in explaining this scenario for human understanding. But the predators don't weigh their options; rather they react to an instinctual response instantly.

Many believe that the engagement of *fear* will cause a *fearful* retreat, such as a wolf tucking his tail and running away, not so. That is not within that *species' behavioral characteristics.*

The actual event would probably look like a standoff, with possibly some barking and very small charges to test the strength of the *defensive posture. The wolf* may even try to engage flight by biting, but through training we have instilled that *draw of the eye.* That *draw of the eye* connects to an instinctual response within the wolf. If the stock were trained well and remained steadfast, the Wolf would simply wander away.

That rekindled faded instinctual response from our stock would be sought in any predatory encounter or pressurized situation. I used wolves because of the wolf's popularity. But any questionable situation or pressure would call up the response of the *Defensive Posture*.

Sometimes one has to just have a laugh, so one of my tricks has been to ask for some assistance in gathering one of my home ranches small fields. So, I just ride away and watch.

The cattle that have been trained with *Predator Awareness* when receiving pressure just go into a *defensive posture.* The harder the push, the tighter they get.

After I've had my chuckle, I ride in and start the cattle; the most common response is, "Boy, those cows sure know you". But, in actuality, it's the *Defensive Posture. Creating movement with the reverse parallel rather than simply driving the stock.*

Individualization

One of the Challenges has been the natural behavioral tendency of a cow and her calf to want to individualize themselves from each other. This seems to occur at approximately three months of age.

That independence creates individualization. Individualization is a weakness for deterring predators and preventing the chase sequence.

For this specific reason, I also deter presence. I have some Ranchers who have chosen this method as their only solution.

I call this method *Scenting*. I have combined several of life's experiences to come up with this deterring effort. One is from my brief experience of being a trapper.

I was taught this trade by an experienced retired Federal Trapper. I did this for two years and can honestly say I didn't enjoy it.

I was managing a ranch in Central Oregon, and the ranch was losing stock. The culprits were Coyotes and Feral dogs.

This trapper occasionally helped on the ranch, but he was enjoying his retirement. When he told me that he would educate me on the trapping skills, I readily accepted. What I soon became aware of is that trapping is a time-consuming occupation.

Through my apprenticeship I quickly realized that it's easier to foul a trap set than set an effective one. Predators are elusive, and I would say that the elusiveness is extreme. That elusiveness makes them sensitive to anything that is not within their normal encounters.

The *predator's awareness* of these unfamiliar stimuli and their elusive nature gave an understanding of the *basic concept* of *deterring with a scent deterrent.*

I began using this method very early on, long before the

Wolf ever became an issue. Combining my experiences into establishing my scenting efforts today.

Before moving to Siskiyou County, I was renting foothill grazing grounds in the country that most Ranchers didn't want, simply because of the ranch's affordability and those ranches' difficulty. Those Ranches were in Central California and on the Central Coast. I was spread out over Mariposa, Tuolumne, Calaveras, Amador, El Dorado, Stanislaus, San Joaquin and San Benito Counties.

Some of you who may have experience, my polite way of commenting about our age, may remember the dreaded times when the protection of the Mountain Lion was confronting California with a perception of pending devastation. In truth, the Wolf has been a repeated panic. But being in Lion habitat I was losing stock to lions.

I was asked by my mother to stop by one of her old friends the next time I was up checking Cattle. My mother's friend was from a long time California ranching family. It happened that I found a fresh Lion kill that mourning; when I stopped by my mother's friend's house, she could tell I was frustrated. I told her of the kill, and she told me to place deterrents. For lions, she liked Chili Powder. I followed her guidelines and didn't have another Lion kill until moving to Siskiyou County.

I didn't continue my deterrent when reaching Siskiyou County, as I didn't see many signs, and it wasn't as steep and brushy as I was used to; it didn't seem to be Lion Country.

The Scenting really came back into play when the wolf OR7 crossed my Ranch. I had become complacent in tolerating my losses, but the Wolf's presence elevated my concerns.

Scenting from a Mustang Colt

(Placing a scent deterrent high to avoid a predator's contact, contact would equal understanding.)

In my research I made quite a bit of effort examining the

effectiveness of the Range Rider scenario. The efforts of the Range Rider are to keep and move the stock in a herd group, and while doing this, you leave the scent of human presence.

The response

What I quickly realized is the action of human presence certainly heightened the predator's elusiveness but didn't prevent presence. Tracks and scat were still found. Whereas scenting properly engages *fear* and *fear's* responses rather than just elevating the predator's elusiveness.

The deterrent must fit

Predator Awareness is about the same deterring effort. Rather than using labor to keep the stock together, I rekindle a faded instinctual response. The scent of human presence left by the Range Riders is then replaced with the application of non-understandable concentrated manufactured scent.

Our difficulty here is our stock's water supply. One can only keep cattle grouped if there is sufficient water. Cattle that are kept off water don't do well. Cattle will survive short feed, but shortening their water supply is rapidly detrimental.

Often, when one thinks of one's herd, the thought goes straight to the total count.

There is a pride in this Industry that's directly related to size and to the years of the ranches established performance. That pride and prestige are real and should be worn as a badge of honor.

But when we are talking about herd size as a deterrent, we need to ask ourselves about our specific range and what is the normal grouping of the stock in that range.

When you make an effort to understand your cattle's use of your ranges, you'll find that most ranges within the high desert don't support your entire inventory as one large herd group.

Most of the time, the entire herd breaks up into smaller, more spread-out groupings. These small herd groups are based on water availability.

These are the size groups the training is targeted at.

Often, I have heard or been asked the question about how you find the time to train all of your stock.

If one realizes that this effort isn't the same as other management practices and establishes that predators are actually a financial concern, then one can weigh out its importance and its suitability to your operation.

A successful outcome for anything requires effort; *Predator Awareness* isn't any different.

Many times, when we think of processing for something such as a preg check, we schedule a vet and a crew on a specific day. We prepare by getting the required stock in position the days before, vaccines delivered, and all of the facilities readied.

This management practice has taken a lot of time and effort. It's planning as well as performance has taken a lot of people.

Predator Awareness training is not this type of event. It's about one person, their dog and the cattle. Your actions are guiding the pseudo-predator and being there to stop its actions when the stock has responded correctly.

This *training session* isn't about gathering into the corrals; it's about *training where the risks occur.* When we're in the corrals, we should concern ourselves to be mindful of the *work with effort* so as not to promote reactionary responses that predators will capitalize on. But the actual *training* should be done where the predator encounters are most likely to happen.

These training events and the required tune-ups are better performed with the least amount of human interaction. This training is intended to instill a response to a predator's presence. We want to keep the human interactions to a minimum, focusing on the stock's reactions to the *pseudo-predator.*

We are instilling a favorable response to a predator's presence. Our efforts, with the training techniques of pressure and release, are establishing that if the cows choose the *Defensive Posture* and the *Standing Solution,* they will find relief from the pressure.

The training of the stock for a response to a *predator's presence* is the effort. But everyday events such as the preg check should be done with an awareness of maintaining the Herd Group.

Controlling the stock's movement through *Stockmanship* rather than that '*Do as I Say* mentality'. Not seeking control threw submission, but understanding our actions, guide their responses.

Providing an engagement of responses for the stock that provides a calm and effective solution to predators.

I believe that getting around the range is still a valuable deterrent. It gives you an opportunity to monitor signs of presence as well as placing new and unfamiliar scent. Presenting that ever-changing unrecognizable o'factory stimulant and deterring presence.

A deterring effort is a constant awareness that predators are opportunists and through our engagements, we can instill a solution for the cattle that will reestablish an instinctual response that will promote an effective solution.

Chapter Ten

Basic Training

When we are instilling a response in our stock, or whether it's working with a young dog or even a young horse. Many times, we enter into a specific session with an objective in mind.

Goals are good, as are objectives, but the animal that we are working with doesn't understand our objectives or our timeline. The animal has no awareness of our plans or intentions, our interactions with them are only a disruption to their daily routine.

If you think about beginning a conversation with someone, and the discussion would be on a technical procedure. We would begin this conversation with some cordial discussion and work into the technical details as we progress.

When we begin our conversation with our stock it's by our introduction. Our introduction and approach should be a zig zag line of approach rather than a straight online of engagement. The zig zag pattern begins the conversation, whereas the straight on method is just creating flight in the preferred direction.

As to the ranch and our livestock, I realize that the Hollywood version is always a rodeo; just step aboard and 'let er buck', "get er done".

Often, when we work with animals, our focus is on the

conclusion (getting done) rather than a conversation. That 'Do as I say mentality' creates *fear* and disrupts an efficient and stress-free effort.

I understand time is valuable and possibly the objective is simply completing the required task in as quick of time as possible. In developing skills that *work with* the animal's *behavioral tendencies* we can minimize our efforts, rather than escalating them.

If the situation is difficult, we should then ask why? It's a tough analysis sometimes, but we are supposed to be the thinkers. A good question is, how can I do this more effectively? Rather than turning the blame on the animals.

If one thinks of our interactions with the animals, one may think of our position as the instructor. As the instructor our response to the animal's efforts could be looked at as grading. That grading of the animal is a response to what we have asked or discouraged.

Our mistake is often our preconceived thoughts of what we want the animal's response to be. Rather than waiting to see what the result is, we conclude a problem and intervene and force our desire because it's just not going fast enough rather than giving a moment in order for the animal to have a successful response.

Many times, our effort to speed things up, rather than abiding by the natural pace, only creates more stress and requires more attempts. The result is drawing out the timeline and raising the stress level.

Such a moment is the realization of whether we ourselves are prepared for what the animals relay back to us.

Effective Pressure and Release training is about recognizing small but positive efforts. The recognition that we present, is the release of that pressure. Intervening because the animal isn't responding fast enough evades the release and simply applies more pressure.

This elevated pressure and stress is using *fear* as a tool for submission.

The Release

Often, I'm asked the question, how long do I release for? I would say it depends on the situation. The *release* is several things; it's a relief to apply pressure. It's also a recognition of that *eye forward question, of what?* The *release* is also a reward for a proper response as well as a good effort.

Actually, the question of how long? Is more directed at the *dwell. The dwell* is about giving the animal enough time to connect what just happened. *The dwell* sets a response to memory. The *dwell is the length of the release.*

The *release* is just the stopping of pressure because the desired result has been achieved.

Going back to the young colt and engaging movement, the

engagement of the movement was a success. One could say that a release could be on that accomplishment. How long would it be for us to realize that we asked the colt to move, and it did, so now I can release. Probably the release would be quicker than this written explanation.

Processing

Another example would be loading the cattle into a crowding alley; let's say your job is loading this crowding alley that leads to the chute where a vet is preg-checking. The vet, as well as the boss want a continuous and quiet flow of the stock. Promoting a low-stress event.

This is definitely the time to understand the *release.*

There are as many different corral arrangements as there are human personalities. But they all work cattle from a group to that individualized status. Often a mistake is pressuring the individualized stock to move alone, encouraging individualization. This is a mistake and encourages individualization by a predator's pressure, as well as an elevated stress for the individualized stock.

Another mistake is addressing a large group. Trying to push multiple animals all at once down the crowding alley. Presenting *rear pressure* that engages that *eye's front pressure point,* which causes the stock to draw their attention to the rear rather than moving forward.

You want to load the stock into this alley in the same position as they will be lined out in that alley. Head to tail in single file, proceeding forward.

You begin this movement with the reverse parallel. This movement is a group movement, not an individualized *chase sequence.*

If there is a pause, that pause is from the eye's forward pressure point, which stops movement.

Determine what created the pause and make the proper adjustment.

By allowing the stock who paused, to return to the herd group and then giving a brief pause, before re-engaging, you have reestablished comfort within the herd group, re-enforcing the *defensive posture.*

Encouraging the group to move by addressing the first individual and addressing the correct pressure point. This pressure point would be the reverse position on our previous diagram. By pressuring this position, with the animal between you and the corral, you create a *lead for the herd group* in the opposite direction of your applied pressure. This is the reverse parallel.

By continuing that movement through the remaining stock, you promote herd action. Lining out and moving forward.

Just as we chatted in gate management about your connection to the *eye and your distance to that eye, that distance monitors speed. That same action is how you control the herd's advancement. By stepping into the eye, you slow or pause movement. Like closing that gate, that we previously chatted about. By stepping away you speed up or tighten the herd posture.*

Giving you, the *stockman, control* in managing the stock's individual position as well as the speed within the lined-out herd group.

This connection to the eye, with the applied *reverse parallel,* will achieve the stock proceeding *forward in a single file.*

Your line of addressing the stock may be a straight line, but you would probably leave a staggered pattern of tracks, much like one who overindulged at the local watering hole. This altered route is intentional and is simply adjusting the applied pressure, just as riding the break we previously chatted about.

Then, by turning and following the group's movement, possibly waving or rattling the flag to encourage the group's continued progress, you address not only the day's work but a positive herd posture.

How long is the applied pressure? It is actually your speed and pace within their flight zone; as you're approaching this individual, you're applying parallel pressure.

A measurement of time would be measured by your pace. The distance would be from your initial address, then proceeding to that chosen individual and creating movement; that pressure and release would be a time measurement of seconds.

Grading

We often interject our opinion of an animal's performance. The animal is certainly not concerned about their grade point average. But our efforts will then progress into seeking a better specific minor result, deviating from the main effort.

Often, we lose sight of the main objective of our session and get entangled in the smaller achievements or discouragements. Simply pecking and scratching like a hen looking for something, but never allowing a release. Constantly picking and crowding the stock in an effort to speed things up.

This *grading* of every response loses sight of the building of the foundation of a working relationship.

The focus on the animal becomes our reward or rebuke. Both establish responses of seeking their reward or fearing our negative response. Taking the animal's attention away from the task at hand and rather focusing it on our approval or discouragement of every moment.

Many times, we try to engage a session as a multi-task

event. When we make time for training, that training should be the main objective.

Training *predator awareness* isn't something that can be intertwined with other management tasks. Although tactics of *poor Stockmanship* can disrupt our efforts of discouraging a predator.

Just as you wouldn't expect a predator to show up at your corrals while you are preg-checking, that management task has no correlation to the objective of deterring a predator's presence. Neither does anything like low-stress Stockmanship. It has no trainable correlation for our cattle into connecting reactions to a predator's applied pressure in the *natural world.*

Low stress stockmanship avoids a stressful management encounter. By understanding our actions, we affect the stock's responses. By not encouraging a low-stress event, we enter into that "*do as I say method*" and aid the predator's objective of individualization.

Predators will only promote an elevated stress level. It's our Stockmanship that can shape a beneficial response to such pressures.

A predator's pressure is a separate occurrence. But if we are aware that *low stress stockmanship* is effective for our management practices. The training of *predator awareness can have the same benefits towards a predator's presence.* Realizing each session has its purpose puts those priorities in place. We must also realize *poor Stockmanship disrupts* any interaction and only benefits the predators.

Effort

As Instructors we need to recognize effort rather than grading each specific action.

As to pups, allow mistakes rather than interrupting an effort. The effort won't be perfect, but the effort itself is. You can refine and control actions after you have established a purpose. The purpose of the pup's training is the handling of our stock. The pup's efforts won't match a seasoned stock dog, though comparing the two isn't a just comparison.

A pup's training starts as early as possible. The '*listen up*' is the first building block.

At a very early age, a whistle or call, which gets the pup to direct its attention to you, the trainer or instructor, that call for attention is the 'listen up'.

Separate and individualized this youngster, so it's just you and them going on this adventure; when I do this, I carry the pup. I find a place that's open, put the pup down and let him explore. He will be very interested in exploring. Pretty soon, he'll realize he is lost, and all alone, often there will be a little whine. At this time, you present the '*listen up*'.

The pup will look to where that came from; after he looks at you, repeat the call, then walk over and pick him up.

This whistle or call, which is *the listen-up,* gets the pup to direct its attention to you, the trainer or the instructor. That recognition then requires a simple reward, with that initial reward include name recognition, Sam, good.

Young pups don't necessarily recognize their names, but it's as good of a time as any to begin name recognition.

This '*listen up*' will be a commonly repeated action. Its intention is to get your dog's attention whenever you request it.

The '*listen up*' is not a come here call or an action that must be followed by a command. But, as time goes on, the 'listen up' will become a direction change or a pause.

Often, young dogs are too aggressive or will chase; either action should just be interrupted and redirected rather than being stopped and reprimanded.

The simple '*listen up*' is intended to interrupt an action with a positive outcome. The '*listen up*' is a building block that shouldn't be under-practice or overlooked.

Rewards or admonishments should be kept to simple acknowledgements rather than abundant affection or physical punishment.

The acknowledgment for a good job should be "good". As to a not-so-good effort, I use the dog's name sternly or simply a stern ahhh.

If the 'listen up' is not acknowledged or doesn't interrupt the pup's action, then you need to work on that specifically.

After the '*listen up',* the 'recall' or 'come here' is established.

The reward or rebuke becomes pointless to us as well as to the pup. The only thing that matters is the accomplishment of the task at hand. The task at hand with the young dog is the effort.

The send will be the next building block and adding a hold or stop and then direction.

Many handlers or Trainers install a 'come here' or 'get behind!' for younger dogs rather than the '*listen up'.* In the future development of that pup, it keeps the dog in closer contact, fearing losing the connection to the trainer or handler. Often when you're performing a task, let's say a gather in some rugged country.

You'll send your dog out after some renegade stock and proceed to follow; if you have worked on that 'come here' method extensively and other riders are helping on this day, rather than sticking to the task at hand, the dog may seek another riders company because it finds comfort and relief there, rather than tending to the task at hand, the wayward stock.

This 'get behind' training can also weaken the send when that send's destination is out of sight of the handler. Often times, in brushy or steep country, the trainers follow will be at a much slower pace because of personal safety. Whereas the dog's responsibility is stopping and controlling the stock within this rugged terrain and holding that group until the handler arrives and decides what needs to be accomplished.

This action is in itself, if used prior to the turnout, is the basic training session of the *defensive posture of the herd group* and *the Standing Solution* and will eliminate the *chase sequence* we just spoke of. Making gathering and finding the stock an easier and more enjoyable event.

The recall or 'come here' or 'get behind' weakens the distance of future performance.

Often, the reasoning of the command is because of the dog's action. The dog is doing something that the trainer is not fond of.

The command stops the action, and by 'getting behind', the trainer's pressure towards the dog stops.

The conclusion for the dog is not a training action, but an action to stop effort and tell the dog the correct and comfortable place for you is behind my horse. Promoting the quit, rather than encouraging the action.

The command of 'get behind' actually interrupts an action.

The *'listen up'* intervenes without discouraging effort but simply pauses it. I need to clarify the action of 'get behind' it is a necessary skill; an example would be an early morning gather when the dew on the grass relays the cattle's location. At this time, a dog in front of the riders will simply create more work in looking for cattle that don't exist by the trails of the disturbed dew. The 'get behind' does have a purpose; we need to understand that it's not for all responses to every action. Its use shouldn't be a disengagement of an unfavorable action. Often, the 'get behind' replaces the use of a choke collar and the command of 'heel'.

Young horses

When we're talking about young horses, the '*listen up'* is a pressure point pressure, that is the disengagement of the youngster's hind quarters. This begins as early as possible, preferably when you are beginning halter training. I prefer a flag to engage that inside bending of the youngster's entire body. The flag is held downward.

If you're bending the youngster to the left, you're disengaging the hind quarters to the right. In your left hand will be the lead, and in your right, the flag. As you direct the youngster's nose to the left, move the flag, not touching the youngster and be aware of a response from that left hind leg, such as a kick. As this youngster progresses, I will engage the youngster at a standstill using the handle of the flag. I will apply direct pressure in the area where my heel would be if I were mounted.

The pressure would be a soft, steady applied pressure. Not a jabbing or poking action. When the youngster steps away, stop. This is preparing the youngster for your applied heel pressure.

If the youngster releases its hind quarters, you accomplished the task, praise the youngster and quit for the day.

This begins an early response to one of the most important pressure points in good Horsemanship. Developing a response that will be used for this individual's lifetime.

By instilling this disengagement of the hind quarters and practicing it in both directions, you create a safe and controlled response to the young horse's reactions to a fearful situation. By disengaging the hind quarters, you remove a horse's power and its ability to buck or run off.

One could realize that this '*listen up*' is a panic button or 'safe zone' for the rider.

By establishing a disengagement of the right and left side, you also help and aid your confidence as a rider.

By practicing this skill, left-hand and left-leg pressure it will arc the horse's body, and the same effort will be engaged on the right side.

Preparing and training this skill prior to the actual need, keeps any unfortunate event from turning into a bad memory.

This training of yielding to mid-body pressure establishes the foundation for many other responses, such as a side pass, opening and closing a gate, yielding to a rope when dragging a calf to the branding fire as well as many more efforts.

The Stocks "listen up"

As to our stock *the listen up* is an *awareness* of anything within the stocks proximity. The action it's self, is the *stop.*

Initially this group is of mixed ages; calves should be at least a month of age. We're just beginning developing their *herd awareness* and working towards a *defensive herd group.* We spoke a bit earlier about *binding pairs. H*opefully, your efforts there have been accomplished.

But if it wasn't accomplished and the cows are not accustomed to dogs, that will be the first effort. I call this *predatory desensitization;* this action is movement outside of their flight zone and slowly progressing within that zone.

Predatory Desensitization removes movement. This action of movement and non-pressure, as well as retreats or simply walking away, is a perception of noninterest towards the stock. That perception is from the stock and about the dog's presence.

This desensitization may require multiple sessions, depending on the stock's sensitivity.

After one completes *predatory desensitization,* one should begin building the *herd group.* That is done with a slow and steady *lateral movement* outside of the flight zone, slowly encroaching into the zone.

A lone fast-moving *pseudo-predator* helps in developing the herd posture. It's not a defensive posture, but it begins an *awareness* that will be built on.

This is the basic effort, but I still call the actual stop as being that simple '*listen up'*.

But the development of the herd is required prior to the engagement of the stop. The action of the stop is when the stock resembles a *herd group. Stop the cattle, with each session the herd posture will become a tighter grouping.*

The Down

Back to the pups, once they're traveling with you, for me, I mean on the range following my horse. The next skill will be a down; this is still for a pup under a year of age.

Actually, the pups tell me when they are requiring a down. When the '*listen up'* begins stretching its boundaries. Some pups require it early, and some never do. What I mean about the '*listen up*' stretching out, is when the whistle or command falls on deaf ears. Often this is when the youngster realizes, this is fun, rather than the effort being a function of purpose.

The down command will be a contact training. The pup will go on a lead, I use a lariat. I always pack a rope. The training, won't be done on horseback but afoot.

At the beginning of this training, the pup isn't going to understand, and *fear* will show up. This is an all-new interaction; you have ahold of him.

This first session is about just walking on a leash. Once the youngster is timidly in control and yielding to pressure, I like a stop to be a sit.

If you accomplish these tasks, praise and release, meaning this session is complete. When I say complete, I just mean that specific item. For me, that will just be turning the pup loose and riding on. That little session was a rest for my horse and around twenty minutes with the pup.

Remember that the dwell and release are as important as the training action. In the next session, I will use a staff on the line and walking. I will ask the young pup to move away in a direction, just as with the colt and the training of releasing its hind quarters.

The reason I don't use a flag on pups is that young dogs will focus and sometimes play with the flag, taking their mind off the objective of learning my commands and simply getting excited and playing.

With the staff or cane, I can touch the pups or tap their hindquarters to engage in movement without engaging in a playful response.

I will give a signal for sending, away or bye. For me, those are whistles. This addresses direction as well as the send. Try to keep your commands consistent.

This will take multiple sessions to get the confidence and establish the control you desire. When you are back in the field, you can present situations that gauge your pup's awareness of your instructions. You will then recognize the need for additional work or realize you can move forward.

Most often, I start this youngster with a lead dog. That's

usually an older dog who understands my commands.

When it comes to the purpose of the effort, the down. I have worked in, leash training and began movement and set a base for the direction of the send. These have been engaged with the aid of the cane or the staff, giving the youngster an understanding of my requests from the positions of presentation.

All done with the training skill of Pressure and release.

So, the pup is sent on the line, basically in a circle in either direction, in a tight enough lead that I can still touch the pup with the staff. When the pup is traveling well, I will quietly switch the staff to the lead hand, give the '*listen up*' command and block the youngster's forward movement simply by blocking it with the staff.

This is in order to strengthen and re-establish the '*listen up'* command. This is an important command. I will grade the individual, not critiquing the youngster but seeing his needs for added sessions or moving on.

When the pup's commands are satisfactory, I will then begin the down. The pause or the stop, or the '*listen up'* has been re-established and strengthened. In that pause, I will give the 'down command'. Blocking directional movement and pulling its lead while tapping the ground in front of the pup and repeating the 'down command '.

The response will be confusion; in the confusion, the pup will try to escape the situation. Keeping a firm hold on the lead, backing up, and preventing an escape to either side, the pup will soon lay sternum down. Immediately release all pressure and praise, Sam, good.

I engage in movement again because I don't want the down to become a quitting signal. I won't constantly use this command, but I will use it lightly to ensure its presence and understanding.

What the training of the down did was to strengthen the '*listen up*'. What the down is, in actuality, is a time out for applied pressures.

Chapter Eleven

Symbiotic Relationships

An observation that I have been witness to, is the many symbiotic relationships in nature. Just as flowers are a joy to look at, they are part of a bee's life. One without the other doesn't fair well. We often concentrate on one piece of the puzzle such as admiring the flowers but spraying the bugs that bother us as we are enjoying nature.

One commonly observed relationship to predators is that specific species of birds seem to follow the predators. The thought is they feed on the carrion after the predator has completed its depredation. The species of birds include Vultures, Ravens, Magpies and Jays. All of these birds are common at a depredation sight. The bird's actions can actually degrade the evidence in the investigation of a depredation. Most of the investigating agency's request covering the carcass with a tarp so the birds don't jeopardize the evidence of the investigation. These birds have a symbiotic relationship with predators. Many people's observations are that they follow the predators.

My observations appear to conclude differently. It may be that it's more of a partnership rather than one following the other. Possibly, it's an effort to seek opportunities which benefit each species.

While changing irrigation water one evening I noticed a calf who appeared off. So, after checking it, it definitely needed to be doctored as it had summer pneumonia. The time of day would not allow time for me to saddle up and take care of this calf that evening. The next day, I was already scheduled for the first half of the day. This calf would need to wait until the next afternoon for it's needed attention, a definite gamble.

The next afternoon, I arrived at the pasture and was disheartened to see an overwhelming presence of birds. I have not seen this many birds all summer. My conclusion was that I was too late in my attempt to save the calf. But an odd fact was evident: the buzzards were circling like some old Western movie. The magpies and ravens were all in their respective groups but all in different areas. There were not any birds concentrated on a specific sight.

Ravens tend to frustrate me with their constant noise. So, as one flew by mocking me, I discharged a round in its direction. There was total silence, and instantly, all of the birds flew away.

As I rode out to find the presumed dead calf, there it lay, still alive, so I happily doctored it and put it together with its mother, brought some other cows closer and left the calf to get better within this protective herd group.

The result was the next mourning, the calf was chewing its cud and got up and made an outstandingly big stretch. For the readers who aren't aware those actions are as good as it gets.

In reflection, the birds proceeded a predator's presence. The calf would have been an easy prey victim for almost any predator. The birds themselves would have been satisfied due to natural circumstances in a rather short time. The fact is that the birds were very attuned in detecting the sick calf prior to any predication or predators' presence.

What if the birds do not follow the predators but rather lead the predators to an opportunity? With that thought it may present an answer to why the wolves are being attracted to bone yards. I realize that boneyards are a legitimate attractant, but are the birds just a neon sign alerting the wolves to the bone yards locations?

By the realization that birds may scout our herds for opportunities. Can that knowledge assist us in monitoring our herd's health? Not waiting for the abundant concentration of birds but the awareness of birds' presence as searching for opportunities. When any bird presence is noticed it may suggest further investigation efforts are required.

In monitoring for others on forestry permits, we occasionally do some cowboying with the owners. We were moving some cattle to fresh feed; I asked the owner if they noticed anything. Their response was, "Not really." When I mentioned the lack of birds, especially magpies and ravens, they then readily noticed their absence.

Scenting moves predators; I have come to believe that when you deter predators, those symbiotic relationships leave as well.

Going back to the sick calf, another possible answer for the lack of predatory pressures is that I had my first mammal depredation in eight years at this same location approximately 50 days prior. My conclusion at that time was that I had been neglectful in my duties of applying deterrents. My scenting diligence and deterrent placement certainly had been elevated since.

That elevated deterring effort towards the predators may have kept the predators at bay. The fact that the birds preceded the predators may support that Hollywood version of the buzzards circling, suggesting a bleak outcome to the situation.

Suggesting that the birds may often act as an alert to opportunities for the predators.

From previous experiences along with this one, I am confident that scenting is a viable deterrent.

Realizing that predators do not always require an animal that is off its game. The predator can certainly subdue healthy animals, but all predators are opportunistic. I believe that a predator will continue to compound any opportunity to its own benefit. But recognizing clues of presence and not presenting opportunities may be a small step in the right direction of preventing a conflict.

Realizing animal or even avian behavior, aids our efforts in deterring a predator's presence. It also enables us to create deterrents that will effectively ignite the instinct of self-preservation. Instinctually engaging a response which tells the predator to move on.

After doctoring the calf, one can assist the sick calf by placing it with other cattle and settling those cattle into a herd group. Also, by scenting the area around the calf, you present a perimeter deterrent to a predator's presence. Allowing the extra protection and security for the calf so it may have adequate time to recover from its illness.

Bear / Wolf relationship.

Another possible relationship could be between black bears and wolves. Both are apex predators, but they don't seem to be opponents. Often, the two tracks are found in relatively close proximity. There are often photographs showing the two species together in camera traps. As far as other apex predators, such as lions, those situations are not observed. The coyotes appear to be solely with coyotes. The relationship between bears and wolves may have something to do with scat. The bears being omnivores, the cellulosic scats are often found, but the meat scat seems to disappear quickly. Presenting a theory that the meat scats are part of the food chain. The thought is to place a trail camera at those locations.

Usually, by the time I return with a camera, the scat has already performed it's disappearing act.

It does appear that when wolves are in an area, coyotes seem to relocate to another area void of wolf presence. When the wolves move to a new location, the coyotes seem to return. Our home ranch borders the Lower Klamath Wildlife Refuge, and we normally have an abundant population of coyotes. Monitoring coyotes has been advantageous in determining the possibility of wolf presence.

Certain birds are predators

As birds can be strictly an effect on carrion, some birds are predators themselves. This is a baby fawn, and upon inspection, it appeared to have no broken bones from something such as a strike from an automobile. Within a quarter of a mile of its location, there was another fawn in the same condition. Also present was another eagle and more buzzards. My conclusion in this observation is that the doe was lost most likely in a roadkill scenario. Leaving the two fawns unprotected and vulnerable.

I have personally witnessed eagle depredations. We often think of a depredation as a battle, but often they are not. This particular depredation was a bald eagle on a two-day-old beef calf. The calf had recently nursed and had laid down for its nap. As the mother came to the hay wagon for her mourning feeding, she left the calf unprotected.

An eagle that was perched in a nearby tree flew down to the calf and landed on it. The landing and take-off appeared to be all one continuous motion. As the eagle flew off to the top of a tree, the calf jumped up as if startled and jumped away. As I watched the calf, it was up and down several times and then finally laying down to die from the injuries it had received. The eagle had pierced the calf with its talons and proceeded to a tree, waiting for the depredation to reach its conclusion. Definitely not much of a battle.

I immediately went to the calf and picked it up. The intent was to graft another calf onto the mother, mitigating the loss. When grafting an adoption onto the mother, we require the dead calf's hide, which we tie onto the new calf, fooling the mother into believing that all is well.

Upon skinning the calf, the wounds were apparent but didn't appear to be lethal. So, upon opening the thorax cavity, the blood attested to the depth of penetration that the eagle's talons had achieved.

I have had questions from people about deterring avian predators. What I have found is that the defensive posture of the herd group seems to effectively deter avian predications as well.

My location in the Klamath Basin is home to the largest seasonal eagle populations in the lower forty-eight. So, efforts to deter mammal predators seem to have helped in deterring other predications. An unintended bonus to deterring a specific predator, the wolf.

By deterring mammal predators, avian presence seems to be reduced. But it does not seem to deter other birds, such as songbirds, from using the cattle as foot warmers. But the birds seeking warmth seem to be the only avian pressure to the herd group.

Chapter Twelve

Monitoring

When we're gauging our effectiveness of predator awareness, monitoring affords us several opportunities. If a scenting program is in place, it affords one the opportunity to freshen such a deterrent.

It's important to present a new scent to avoid desensitization.

Desensitization happens with repetition. Changing from a fabric softener to a cleaning agent is an example that prevents the predators from becoming comfortable and desensitized.

The re-application of scent should be on a seven-to-ten-day schedule. Each individual application should be applied to cover no more than five square miles. In areas that are known to have travel routes and the intention is to detour that route, closer staggered placements will be needed.

A Virtual Scent Fence

When we present this detour to a predator, it's intended to alter their travel patterns. One could think of such a placement being a *virtual scent fence.* Placed for security rather than containment.

As a rancher, we think of a physical fence, for such things as containing our livestock within specific boundaries. This *virtual scent fence* is placed to discourage a predator's presence, detouring its intended route. By presenting an unknown, we limit the predators' options.

When placing a *scent fence* your placing it across the predator's migratory trail, the placements should be approximately a hundred yards from each other. Those closer placements should cross the road or trail that you suspect is the corridor. The total needed placements would require two or three deterrents. I would also place another scent deterrent further down that trail, at about a quarter mile, in a parallel placement to that trail.

The *scent fence* may possibly be outside of our boundaries. But, from our understanding of the terrain and the known predator's travel routes we may decide this is a good spot to create a detour. Rerouting a predator's intended destination, which provides security.

By the placement of new and different scents, such a change promotes a non-understanding. Non-understanding promotes an instinctual response, a reaction rather than an action.

Don't re-scent previously placed deterrents; present the new deterrent about a half to one-mile distance from the previously placed scent.

Leave the previous deterrent in place so a rain event can reactivate it.

Where?

When you're placing a scent deterrent, there are certain locations that are more effective than others. Often, the scent hanger itself is as much of a requirement for the placement as anything.

It needs to be high enough to avoid contact, I prefer at least six feet off of the ground. I often scent by horseback, easily accomplishing the height requirement, but have placed others from the bed of my pickup.

The presentation is important, but the most important is not allowing the predators to understand the placement. The predators contacting the scents presentation would provide them with understanding. That understanding would negate the effort.

Game trails are the most effective, but cow trails often serve the cattle as well as the predators. Either cow or game trails always have a destination, just as our roads do. Most often, the trail is linked to water and feed.

As with all species, water is the key to life. So, in limited conditions such as high desert, the trails leading to the water sources are a priority. Placements there may need to present a *scent fence* and those placements would be tighter. When a game trail crosses over a man-made paved road, such an intersection can be a good location and an opportunity to place a deterrent to re-direct the predators.

Ridge lines are often a broad trail rather than a defined one, but placements on those ridges are effective.

Presentation and Placement

How we present the deterrent is also important, what we are presenting is an un-resolvable question. So, it can't be linked to a scent that's understandable.

Those scents which are common in the Natural Environment, such as green wood or a treated post, are understandable. The scent must be and remain an unknown, as well as a foreign and strange occurrence to the predator. Don't apply the scent or allow contact with an understandable and common scent. The effort in the presentation is creating an unknown; that unknown factor is the deterrent in itself.

(Deadwood offers a good scenting location. Dribble the scent as high as possible to avoid contact, which would equate to understanding.)

It must be placed high and in the breeze to remain an unsolvable mystery.

Do Not pour it on the ground

Humans and their accompanied scents are completely understandable and quickly engage a cautionary response for the predators. I realize many say that human presence is a deterrent. Human presence only heightens the elusive nature of the predator rather than engaging an instinctual response.

Predators are comfortable within close proximity to mankind; mankind offers many opportunities.

My point is if it's *understandable,* that equals caution for the predators. *Non-understanding* brings forward an *instinctual response: fear.*

Presentation

I like to saturate an old sock (a wick) stuff into another and then tie it around a branch. When I saturate the wick, I do it in a container, just enough to make it wet but not dripping.

During transportation, I prefer to keep the saturated socks in a container, like a Tupperware container. Not allowing the scent to be spread until I choose its deterring location. This strengthens its shock factor and doesn't allow a slow, progressive introduction of this unknown. Presenting that *boogie man event.*

We are trying to present an unknown, for that reason choose

hanger socks that are dark, like black or brown.

I realize many don't realize why some cowboys wear white shirts. They're cool through the summer and they're a deterrent to pests, such as mosquitoes, green heads and deer flies. But they're also better than wearing an orange vest. A rider wearing a white shirt is easily spotted from quite a distance. White stands out.

Sometimes one hears of wildlife encounters that surprised the wildlife. The white shirt let's wildlife or even other humans know your location.

Don't use white socks as hangers.

Visualization

Monitoring also affords us the opportunity to look around for signs that attest to a predator's presence.

Certainly, tracks would be one of those signs, but as the ranges are seasonal, often in our area, the tracking conditions aren't so good. Fortunately, I'm riding and have my *pseudo predators* with me for some possibilities of a training tune-up of the cattle.

I have stated that contact with predators only affords them an opportunity for understanding. For that reason, I place my scents at midday, when contact is highly unlikely.

Unlikely, but not out of the realms of possibility. In my time of practicing this afternoon observation, I confronted three wolves by surprise. One was on a one-hundred-degree afternoon, and the black female was lying in a stream, escaping the days heat.

The other was a gray male and it had three deer in a defensive group. Deciding which one?

The other was patrolling my stock here on the home ranch; it was a collared wolf. I had been busy and had not scented here for a while, attesting to the importance of keeping up with the schedule.

This wolf had a hair coat like a Coyote. I had been told by many that wolves and coyotes would not crossbreed. I'm not so sure; I believe in instinct. If *self-preservation* is the strongest, *procreation* is a close second. Recently, I have seen some coyotes with abnormally large feet.

When I'm riding, and I have my *pseudo-predators* with me, I prefer that they travel out in front of me. It doesn't mean they are running amuck; I have the ability to control them at anytime. But, as I just attested too, tracking can be a challenge when you're in high desert conditions.

The dog's actions are the same as my purpose for being out here. I'm placing a deterrent that will create a response in the predator's O'factory system. The *pseudo-predators* also respond to scents left by the predators. By watching them they'll inform me of a scent left by predators and their previous presence. I have always had dogs; I guess I'm comfortable with them. Through my time with them, I have come to realize some behavioral characteristics that definitely affect performance.

Individual dogs seek and require support; often, with Stock-dogs, it relays to a sendable distance. Often, a young dog, after it's been sent will stop and look back to the handler for reassurance to continue on.

Two dogs will compete and pair up, not abiding to commands but competing against the other dog.

Three dogs, the *Trinity,* is almost a magical occurrence. Their individual weakness vanishes, and their competition with the other individual dogs subsides. The attention to the direction of the handler elevates as if they're in a competition for the handler's approval. Which certainly elevates their performance.

I have found that an odd number of dogs suit performance. Whereas an even number promotes individual competition. Occasionally, I may use five, but my preferred canine pack is three.

When this *Trinity* is out in front, they're all individualized, each dog working an area. In my younger years of hunting pheasants and quail, this *Trinity* was just as admirable.

The *pseudo-predators* will attest to signs that I would have certainly overlooked.

Their ability to detect tracks, urine, or scats is a valuable tool in determining a predator's presence. After some time, their actions become understandable, whether they are testifying to the sex or the species.

When we're talking about the sex of a predator, males mark items such as bushes or trees. Whereas females will mark in the open. The pseudo-predators will show an almost excited action as all three will converge to the location.

Scat

I realize it's not very fashionable to step in scat, but if you're fortunate enough to find it, it should be investigated, as it attests to what the predator has been eating.

I prefer to step in it and smear it by drawing my foot backwards. Smearing and spreading it out. The thin spread makes determining its contents more easily diagnosed.

Hair is not digestible; it remains in the scat. Black hair would be probable to be from a black bovine. Whereas tan and black tipped hair would be from a deer.

The black hair requires further investigation, whereas the deer doesn't. Hair in scats doesn't qualify that a depredation has taken place, only that the predator fed on the carcass. The carcass should be found and investigated as to the reason for the death.

Scats have species characteristics. Canines would have tapered and pinched ends to each segment. Whereas a Bears segment would be blunt. Felines will bury theirs, often in a loose type of soil. The felines are a little more difficult to find.

Their scat would resemble a canine.

Being aware of small piles or mounds is the best way to describe a lion's placement.

There is a difference in species' *behavioral characteristics,* even in *scats.* Felines seem to be stealthier in their actions. That elusiveness even could be related to hiding their scat or simply burying it, whereas Canines present it as a billboard for all to see.

Attractants

We often hear the suggestions of not leaving such a stimulus, such as bone yards or other attractants. I realize that and agree with that statement, but I would also like to point out that the boneyards are foremost in our minds. They're a location that no one admires.

Acknowledging that the bone yards are an attractant, I believe there may be many other attractants on the ranch, from our household trash to our pets. One must also realize that the stock itself may in fact, be the most desirable attractant.

A predator being a predator means that it is constantly seeking opportunity, and searching for opportunity will eventually result in its presence, even if there is no attraction guiding it.

If you're familiar with dogs, it's not an uncommon occurrence for canines to seek or tip over trash cans looking for those discarded treasures. I believe it's a *behavioral characteristic of the canine species. Searching for opportunity.*

Another *stimulus* that would be a draw is an abattoir; in our area, such an attractant has a close proximity to other ranchers. Wether it was the Shasta pack or the Whaleback pack or the young migratory individuals leaving the Rogue pack on their journey of searching for a new home area.

Being in the abattoir's proximity certainly puts a ranch's

stock in the areas of such an attractant.

That fact may question and override other attractants. Discounting the effectiveness and expense of cleaning up those established piles.

Previously, I spoke about a predator's sensitivity and their awareness of things that are outside of their understanding as extreme.

In my trapping experiences, the fact that it's easier to foul a trap set than set an effective one, lends to the conclusion that draws and attractants can be effectively overridden.

In my experience and being adjacent to Ranches suffering predatory losses, an observation in placing scents as deterrents, is that avian presence seems to vanish once I applied the deterrent. My thought is that they followed the exit of the predators.

If an avian presence occurred within the scent's reapplication time frame, I would then mark my previously applied scent off of my effective scent list and not place it again.

I attribute the avian presence as predator presence. Not because of the birds themselves, but as too the symbiotic relationship between those avian species and the predators.

Testing

Another opportunity that monitoring affords is the time to check your cattle's response to pressure.

Most often, their response will be *herd awareness,* moving towards other stock.

That response is the extent of the session. If the cattle chase the dogs or flee, that's a sign that requires further investigation. Possibly attesting to predatory pressures.

Most often when we're looking at signs, the cattle's disposition is also on the top of the list. Their tracts and scat, as well as their cleanliness, will attest to their disposition.

If a cow is going downhill on her own with no outside pressure, most often, she'll stop to defecate. Moving her hind legs slightly forward so as not to dirty them.

I call this specific pile of evidence (Standing patties).

If she's going uphill, she'll continue walking, but the spacing will be the length of each stride. I call this (Walking patties).

Both attest to direction, even if the tracks aren't there because of poor tracking conditions. Standing patties mean they were heading downhill. Walking patties means they were traveling uphill.

Then there are signs such as (Running patties) this isn't good. It's because there fleeing and defecating from the response of *fear.*

The cattle's dropping's attests to disposition as well as the direction of movement. One of the confusing signs in distinguishing a *fear* response is an estrus cycle. As a female is in heat, her scat placement can certainly resemble flight. But, if you follow those signs for a distance, you'll often find some tracks to judge cadence. A female in heat will frequently stop. A cow in flight will not.

Markers

Often, in our everyday chores and busy agendas, we overlook these *markers.* They're actually a very common occurrence, but we simply write them off as litter bugs or just some human's lack of respect for the environment.

What these *markers* are often referred to is simply trash. There is an old saying, "one man's trash is another man's treasure". That saying certainly applies to *monitoring.*

That trash certainly tells a story and deserves to be investigated.

These *markers* attest to specific species, canines as well as bears. Lions seem to have a complete disregard for *markers* and are only interested in edible discards.

The lion's weakness is cardboard boxes.

Both bears and canines have an attraction to aluminum. Wether it's cans or plates, aluminum is an attractant to both species.

This knowledge affords us an opportunity, through investigation, to understand the species and possibly the parameters of their domain. Wether it's investigating trash alongside the paved road or trash found in the middle of National Forest lands far from such roads. That trash often tells a story of a predator's interaction with it.

It's not always that disrespectful human we all blame, but often, the items are a sought-after item of interest for the predators.

The most commonly found *marker* is an aluminum can. Bears, when interacting with the can, just mangle the can. The evidence seems to show a bear's action as unrestricted, and it just enjoys crushing the can. Often, bites or punctures on the can, seem random in placement. The can seems to get the same response as bubble rap does for humans: the uncontrollable urge is to give it a mash.

(A bear influence on a marker)

Whereas the canines show a more controlled interaction. The canines seem to measure the can, being careful to center the can in the center of their mouth. The reason that I make this statement is that the cans that I have found throughout different states all have the same dented structure.

It seems to be the characteristic of canines. The distance of the canine teeth seems to be the only distinguishing characteristic between coyotes and wolves.

Both the adult coyote and wolf seem to center the can within they're mouth and then gently pick it up between their tongue and top canine teeth. It seems as if they then push the can out of their mouth with tongue pressure. Sliding the can across their top canines and leaving a very distinctive and measurable marker.

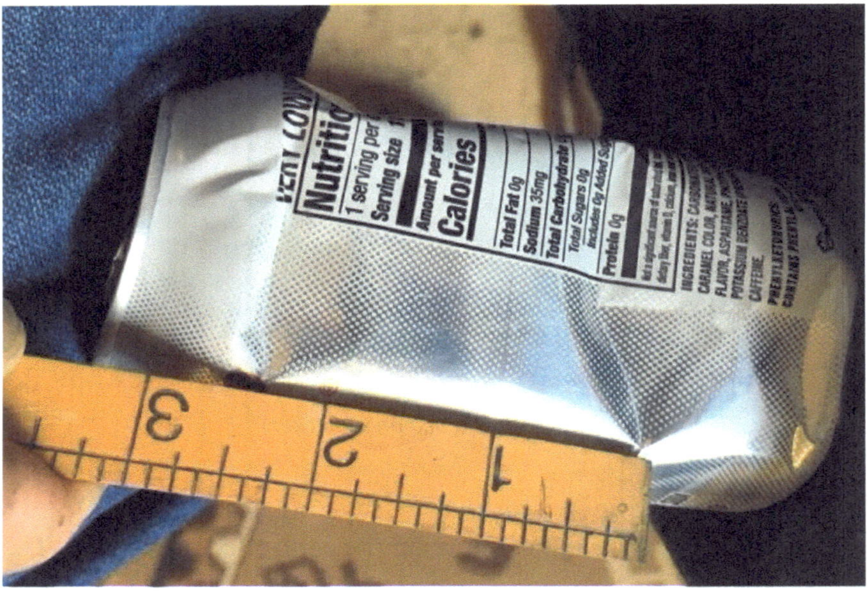

Canine influence

If the canine takes the can to a den sight or a rendezvous sight, the pup's interaction then resembles the interaction much like the bears, simply destroying the can.

The Coyote's measurement will be around two inches. Whereas the wolf's will be around a three-inch measurement.

The can, maybe left where it was found, leaving only the measurable clues, or it may be taken away, possibly as a toy for pups.

The predator's interest in these cans is a useful monitoring tool. When I find and distinguish a game trail, I will look for some trackable soil. I will then leave a new empty aluminum can, an intentionally placed *marker*. Each time I make a monitoring round, it will be one of the things I monitor.

It's been my experience that a predator, if passing by, must engage the marker.

Often leaving the marker as well as their individualized calling card.

If it's gone, I'll find the track and follow it in the direction it went. Remember, I chose the sight for its trackable footing.

If I find a dented can, I'll pick it up and replace it with a new can.

What this placement does, is it helps me detect presence, but also provides clues to the travel patterns of the predators and how they're traveling through this range.

Tracks

I mentioned that tracking is sometimes difficult. The palette that the track is presented in is what allows its form to be held.

Track reading has everything to do with moisture. Adequate moisture means the tracks will hold their shape. In extremely dry conditions, only the type of soil affords an identification.

Often, a soil type that will show an identifiable track just doesn't exist. Disturbed dry soils are a good palette. Vehicle traffic provides a very good fine and loose powder that creates a readable track. Many believe that the roadways are our exclusive domain and are for humans only.

I realize your traveling on the road with a destination in mind and a purpose for getting there. But, when your monitoring, your missing signs if you're not looking.

Stop and look at dirt roads, especially the intersections.

The traffic creates a good palette. The predators will often line out on a fresh tire track. It's an interesting occurrence; with a two-lane dirt road, why does the predator walk on a tire track? I am not sure why; I only know that when it comes to animals, I don't believe in coincidences. Possibly, the disturbed soil is a pleasure for the *predators* to walk in.

Roads are not exclusive to humans; in fact, predators monitor the roads themselves. Humans being humans loose or discard things from their vehicles, simply making the road itself a corridor of opportunity for the predators. The road is routinely used. You can look to any freeway or commuter route in any Urban area, and what you will see lining that roadway is an abundance of billboards filled with advertising.

This phenomenon of advertising is not solely a human characteristic.

The over abundance of those billboards affects us, even if we don't realize it. To make my point, what were the last five billboards you past and what were they advertising? The next question would be, where were they?

The advertising business has desensitized us. I remember the Milk Council's advertising "Milk, it does a body good" with a picture of a gorgeous young lady reclining in some swimwear.

The outrage was intense. Everybody was talking about it. It was the peak of what an advertising agency should hope to achieve.

The point is that in our busy routines, we actually don't notice the billboards.

We have become desensitized to them.

It's not much different for folks in the rural communities; with all the things to do, we're just intent on getting to where we're going. We don't have the time to look at billboards or predatory signs.

Many predators place their billboards; they are placed in the center or on the edge of a well-traveled road. It's the predator's largest sign that an individual predator can place. That billboard is their scat. It states this is mine.

It presents a sign of presence. Such a sign should be looked at for undigested hair. It would also signify the possibility of a game trail crossing. If it's found within a ten-mile distance from your deterring efforts, your efforts should be elevated for a time, just as if you found it on your range.

The placement of the predator's scat is often on the highway or dirt road. Its placement will be in the centre or on the shoulder of the road. Both positions make it possible to prevent it from being run over by vehicles.

Such placement attests to others, including humans, that this road is their domain.

When you think of the roads being the predator's domain, it starts one to think about the many opportunities the roads present.

The engineer that chose the location, chose it because it was the easiest route to a destination.

Humans often look towards tough and formidable country and equate that country to predators. I realize when I say formidable, that's a perception of the individual. A better statement, for a better understanding, maybe ground that's too tough for cattle.

Mount Shasta

There are species that seek such type of country, such as the Rocky Mountain Big Horn sheep. As well as the Mountain Goats.

Their presence there attests to their adaptation to such a rugged environment. Through natural selection, the species has prospered in such rugged terrain. That establishment was due to the pressure of predators; the terrain itself is the relief to predatory pressure.

The predators that engage our stock are there because of the stock. Our stock equals opportunity.

There are many other opportunities than just our stock, but

the fact is those opportunities are closer to mankind rather than further away.

(Antelope)

When you're looking at tough country, it doesn't mean predators won't be there. What it means is they're more likely to be closer to opportunity rather than being further away from it.

Predators will capitalize on any opportunity; mankind presents a long list of those accommodations.

Track Identification

When the wolf was new to this area, I often wondered, is that a wolf track? But, once I saw a wolf track, that question was quickly understandable. The track size answers the question. The size of its track is why so many people put such elevated estimates on the wolf's size.

I imagine that since the predator doesn't hibernate and is an active hunter year-round. Natural selection provided large feet as snowshoes. Allowing a superior advantage in snow or boggy wet soils such as an early spring.

Giving the wolf superior tactical capabilities over their prey.

There is a certain amount of panic about the wolf's presence and the societal turmoil it creates. Often, those conclusions are comprised of fairy tales and childhood stories, establishing a fear of such a terrifying beast.

Its presence conjures up all kinds of scary scenarios. But, as we're focusing on deterrents, one thing we are speaking about is *fear.* As I have said in previous chapters, *understanding one's fear is the first step in overcoming that fear.*

The wolf is certainly a capable threat, but so is a lion or a bear.

My losses prior to the wolf had become an accepted tolerance to such losses.

The wolf societal panic actually helped correct my complacency.

Not only size

A wolf track, as well as a coyote, have characteristics that separate them from other canines.

The front left goes down, the right hind then follows, but the right hind's footfall steps into the front left track. Two footfalls making one track. The right front would reach out, and the left hind would step in that right front print. Creating an unmistakable track recognition for wolves and coyotes. A mono-line track, unique to predators. Every other animal as well as humans, tracks out as a by-linear pattern. A distinctive right and left line.

I have heard many conclusions in identifying tracts, some only attesting to size.

The mature wolf tract is substantial, but so is a mature lion. The distinguishing characteristic is toenails and a by-linear or mono-line pattern. The size is a sign of maturity rather than species. Neither the lion nor the wolf are born with that size of foot; they have to grow into them.

Often sub adults can be as much of a problem as mature animals. Often the mature animals will allow the sub-adults to do the hunting as they lay back and observe.

A lot of tracks

Many times, we find a network of tracks, and we often assume all the tracks were created in one event. It's possible, but a *behavioral characteristic of predators* is the fact of searching for opportunity.

Northern California has gained a new agricultural industry since California legalized *cannabis.*

These new crops are often raided by deer and domestic livestock. These farmers often round up dogs from shelters or from people who are simply ready to downsize. They then simply turn this domestic canine pack loose. The pack's intended purpose is to prevent the stock and the deer from raiding the farmer's product. Simply harassing the intruders, the deer and the livestock.

These domestic packs are an attractant, their lack of understanding of the *natural* world puts them at a disadvantage. Predators, being opportunistic, are soon seeking out such opportunities.

A scenario I often run across is this domestic pack's tracks. Within these domestic canine tracks there will be mixed within, a predator's tracks. Many conclude that the tracks all being together were established at the same time. More than likely, the predator was simply following, looking for an opportunity, not traveling together, but rather a predator stalking prey.

We chatted quite a bit about instinct; I do believe *self-preservation* to be the most powerful. With that said, *procreation* comes in a very close second. This scenario of a domestic canine pack co-mingling with wild wolves affords both of those instincts with plausible opportunities.

A litter from a domestic female that has no *fear* of humans could pass on *behavioral tendencies* of no *fear* to her pups simply by her actions and responses to humans. Rearing a wild, *hi-bred* individual that's accepting of a human's engagement. This hi-bred could be referred to as *fearless* because of his mother's instilled responses to human interactions.

The possibility of a situation getting worse before getting better is certainly a reasonable concern.

Specific to a Wolf

Distinguishing and identifying a track is subject to the pallet that it's presented in. Often, the track itself is inconclusive in its identification, but the line may attest to its creator.

(This track would be from the right front and left hind foot)

(This tract would be from the left front and the right hind)

(This unique picture of the right hind sliding into the left front track is because the wolf looked backwards to its left, disengaging its hips to the right, arching its body and making the footfall outside of its intended placement.

A county road was behind the wolf and was probably the reason for the wolf looking backwards.)

With *Horsemanship,* we establish this same action. This bending of the horse's entire body was earlier described as controlling the youngster's ability to buck or run off. I refer to this as the *listen-up,* the *disengagement* of the hind quarters. Arcing the individual's entire body, the track attests to the wolf's head being turned to the left, looking backwards, disengaging the wolf's right hind leg. Altering the right hind foot's destination.

Chapter Thirteen
Profiteering

Fear equals profit; I realize that a lot of you are scratching your head and wondering, what?

This is going to be an explanation to many that may seem to be about humanity, but it's rather a daily occurrence in life, whether we are applying it to the animal kingdom or our daily routines.

Managing fear isn't a new concept; it's a tried-and-true *practice of control and manipulation.*

One can certainly look at *predators* using *fear* to control a favorable response from their *prey victims* that *benefits* the *predators, the chase.*

One could even say the predators capitalize on that management of fear. When we look towards human interactions, whether local, State, Federal, social or Worldwide, we can certainly see situations that resemble a *predator's* use of *fear* in managing a prey's response. The predatory action can be equated *to the dominating humans* and those humans simply managing another's *fear* for their benefit.

Whether those applied pressures are a suggestion, or to the *natural world,* an actual confrontation, both are about *managing a response to fear.*

When we look towards humanity, we can certainly see such occurrences in the bombardment of daily advertising, suggesting the importance of this product to survive a multitude of pending threats. Wether it's health, security, longevity or anything we require to continue our daily happiness.

All of these efforts are to inform us of the importance of (whatever the product is) and our need for that product.

I understand that some are saying that's just marketing! I agree. Marketing is convincing you that you need their product. That's capitalism; they need you to contribute to their bottom line. I believe that we're on the same line of thought. They have convinced you that you need their product. Was that purchase a manipulation of your *fear* of some impending inevitable occurrence, or simply a product you desired or enjoyed?

The difference is the manipulation of your thoughts and instilling the need for their product.

That difference isn't any different than your daily needs versus your desires, kinda like going to the city and having to decide whether to pay for parking or walk further. Many pay for the parking; that's simply marketing *convenience,* then many that purchase that *convenience* will then purchase an *exercise machine* or enroll in a *jazzercise* class. *Both responses are about marketing.* Managing your decisions for their profits.

The individual that chose to walk, saved his parking fees, as well as getting his or her exercise. But the parking garage is convenient, and many jazzercise classes are full.

When we look to the *political stage,* they not only market *fear but to your perceptions of benefits, as well as promising unattainable goals.* All of this is trying to secure your vote. Politics has become about manipulating *fear. When they successfully manage your fear, they have elevated themselves within society to a position of power. Power affords opportunity, and opportunity results in acquiring profit through the cover of donations.*

When we reach environmental issues, they're intertwined with politics. The environmental regime has become entangled with bureaucracy. As I had stated in a previous chapter, the definition of government is paying someone to manage unproductive results. Environmental issues are about money, for the politician as well as the environmental promotion, whatever it is.

This Worldwide marketing of environmental issues has become the most lucrative segment of the *social society network* to date. Whether you restrict some and embellish others, it's about *profiteering* rather than finding any solutions. I stated in a previous chapter about *anomalies* and their pointless solutions. Any solution results in the same outcome for an anomaly; the true solution is time.

The end of the world has been a marketing ploy for as long as mankind has been in existence. It doesn't matter what society you look back on, that *fear of the impending apocalypse* is there in their recorded history.

True *environmentalism* is just modern societies' *new religion.*

Replacing Sunday's attendance to simply *new social standards* of *recycling* or supporting *clean air marketing restrictions* or whatever the *Green New Deal suggests.*

The passing of the plate to ensure that the church can pay the power bill so the next sermon can have light has changed into the *environmental evangelist* being able to upgrade their jets.

The Sky is Falling

There's an old children's tale of *marketing social panic.* Actually, there are quite a few that show this *societal panic,* but I'm thinking of "Henny Penny in Chicken Little".

Henny Penny was a young hen pecking along in the barnyard when from the limb of the Old Oak tree that she was pecking under fell an acorn, striking Henny on the head. Henny's reply was *the sky is falling.* Soon, the barnyard was all in a panic as *the sky was falling.* This response was an easily accepted reasoning for the barnyard society, as the *sky was falling!*

The marketability of mankind's doom is a *characteristic of humanity.*

Managing Fear isn't a New Concept

I realize that there are many that will contradict any solution, and some more readily than others. Some actually seek confrontation rather than any solution. I often hear those arguments throughout society no matter what the topic is that you're discussing. There are those who simply oppose.

I understand, but if you ponder *fear* and its daily occurrence, you'll begin to realize that it's a *manageable tool. Society* has been showing us its usefulness for years, as well as the predator's tactics of securing their prey. When you begin to think of *fear* in that way, you also can use it and capitalize on it to your benefit. By *managing fear,* we provide our livestock with an alternative response to the *predator's actions.* In turn, the *stock's response* engages the predator's *fear.* By *managing a predator's movement* as well as removing our livestock's *fearful response* we are simply promoting profit.

Fire

Predators and Fire may have some similarities; avoiding a problem may be an easier response than dealing with the problem's actuality.

By managing *fire prevention,* you may circumvent such a fire, which would be much easier than responding to that event.

By instilling management practices that discourage confrontations, you also mitigate those risks.

Both efforts, whether it's fire or predators, are acknowledging and preparing for that just in case.

An effective mitigation of risk

What I have experienced is that the risks that predators present are manageable. The predators are not going to stop what they have been instilled to do: their predators. Social and public endeavors of reestablishing their populations and limiting or eliminating the controls of managing those numbers mean that the task at hand falls to alternative methods.

Understanding behavioral characteristics and realizing that instincts rule over any action puts an effective deterrent within our grasp. Managing an animal's response may seem to be a reach for some, I understand, but a deterring effort doesn't change the predators; it only changes one's own risk.

The realization of *managing risk,* rather than reacting to it, is a reach for some, but that's OK, that's life. Our choices are just that, our choices. The Ranchers, as well as myself, have all seen benefits to these deterring efforts. Whereas a lot of folk's concerns seem to go down that trail of "we have to do something". Repeating that old re-run of reacting rather than preparing.

When you think about applying these methods, you *desire* a proactive result.

Desire is the objective and passion is the effort. Minimal efforts only achieve minimalistic results.

About the Author

I'm Mark Coats, a Rancher who worked his way up through the Ranks of the livestock industry, a rarity within the Ranching world. From cowboying to management and finally ownership. My brief experience as a bureaucrat, strengthened my resolve to endeavor to reach my goals.

Animals have always been a part of my life. My experiences, often put me into remote areas which instilled an awareness to the natural world.

Many mentors have shown me pieces of what I have learned as well as Nature itself.

The natural behavioral characteristics of animals are steadfast responses. Animal's actions don't change with fashion, or with what's popular, nor political correctness. They only have their species behavioral tendencies.

I am not a creator of anything new, I'm only relaying my experiences within the Natural World. These behavioral actions were learned through years of experience. Through my writings, the endeavor is an effort to share my experiences and a way to mitigate predatory risks if those risks are of a concern.

Mark Coats

GLOSSARY

BLM (bureau of land management)

Calvy Cows Cows close to calving "birth."

Dry Cows Non-lactating cows

Pairs A mother cow with her nursing calf

"Pause and Tensity" An instinctual reaction to a non-understandable sensory perception.

"Predatory Desensitization" A term for a Stockmanship Skill of settling cattle. The skill may be found in the first book "Deterring Wolves and Other Predators"

Preg check Checking to see if a cow is bred.

Made in United States
Troutdale, OR
11/03/2026

56845375R00146